Multivariate Analysis for Neuroimaging Data

Atsushi Kawaguchi

Faculty of Medicine, Saga University
Nabeshima, Saga, Japan

CRC Press
Taylor & Francis Gr
Boca Raton   London

CRC Press is an imprint of the
Taylor & Francis Group, an **informa** business

A SCIENCE PUBLISHERS BOOK

T0203647

First edition published 2021
by CRC Press
6000 Broken Sound Parkway NW, Suite 300, Boca Raton, FL 33487-2742

and by CRC Press
4 Park Square, Milton Park, Abingdon, Oxon OX14 4RN

© 2021 Taylor & Francis Group, LLC

CRC Press is an imprint of Taylor & Francis Group, an Informa business

No claim to original U.S. Government works

Reasonable efforts have been made to publish reliable data and information, but the author and publisher cannot assume responsibility for the validity of all materials or the consequences of their use. The authors and publishers have attempted to trace the copyright holders of all material reproduced in this publication and apologize to copyright holders if permission to publish in this form has not been obtained. If any copyright material has not been acknowledged please write and let us know so we may rectify in any future reprint.

Except as permitted under U.S. Copyright Law, no part of this book may be reprinted, reproduced, transmitted, or utilized in any form by any electronic, mechanical, or other means, now known or hereafter invented, including photocopying, microfilming, and recording, or in any information storage or retrieval system, without written permission from the publishers.

For permission to photocopy or use material electronically from this work, access www.copyright.com or contact the Copyright Clearance Center, Inc. (CCC), 222 Rosewood Drive, Danvers, MA 01923, 978-750-8400. For works that are not available on CCC please contact mpkbookspermissions@tandf.co.uk

Trademark notice: Product or corporate names may be trademarks or registered trademarks and are used only for identification and explanation without intent to infringe.

Library of Congress Cataloging-in-Publication Data

Names: Kawaguchi, Atsushi, 1976- author.
Title: Multivariate analysis for neuroimaging data / Atsushi Kawaguchi.
Description: First edition. | Boca Raton : CRC Press ; Taylor & Francis
    Group, 2021. | "CRC Press is an imprint of the Taylor & Francis Group,
    an Informa Business." | Includes bibliographical references and index. |
    Summary: "This book facilitates (statistical) brain imaging data
    analysis. It targets a wide range of researchers interested in
    biostatistics, data science, and neuroscience. It is useful to
    understand the background theory of standard software for neuroimaging
    data analysis. Our state of art multivariate methods developed by our
    research team is explained here using numerical examples that have been
    formulated with the free statistical software R. This book supports the
    application of these methods to your own data in neuroscience and
    encourages new research in biostatistics and data science"-- Provided by
    publisher.
Identifiers: LCCN 2020046688 | ISBN 9780367255329 (hardcover)
Subjects: LCSH: Brain--Imaging--Statistical methods. | Multivariate
    analysis. | R (Computer program language) | Neuroinformatics.
Classification: LCC RC386.6.D52 K39 2021 | DDC 616.8/04754--dc23
LC record available at https://lccn.loc.gov/2020046688

ISBN-13: 978-0-367-25532-9 (hbk)
ISBN-13: 978-0-367-75221-7 (pbk)

Typeset in Times New Roman
by Radiant Productions

Visit the Taylor & Francis Web site at
http://www.taylorandfrancis.com

and the CRC Press Web site at
http://www.crcpress.com

To Machiko, Saito and Junon

Preface

Purpose of the book

This book illustrates methods for statistical brain imaging data analysis from both the perspective of methodology and from the standpoint of application for software implementation in neuroscience research. The illustrated methods include those both commonly used (traditional, standard, or well-established) and state of the art or advanced methods.

Why multivariate analysis

The described method is a multivariate approach originally developed by our research team. Since brain imaging data generally has a highly correlated and complex structure with large amounts of data, categorized into big data, our multivariate approach can be used for dimension reduction by following the application of statistical methods. The R package for most of the methods described is provided to our readers. Understanding the background theory is helpful in implementing the software for original and creative applications and for an unbiased interpretation of the output.

Who this book is for

This book targets a broad audience, such as Ph.D. candidates, graduate students, and researchers in a wide range of disciplines: statistics, biostatistics, psychology, neuroscience, computer science, biology, radiology, psychiatry, and data science.

This book also explains currently developing methods in a conceptual manner, which are useful in further development of a scientist's own research. These methodologies and packages are commonly applied in life science data analysis. Advanced methods to obtain novel insights are introduced, thereby encouraging the development of new methods and applications for research into medicine as a neuroscience.

How to read this book

The commonly used methods are easily implemented owing to established software, which is frequently used among neuroscience researchers. For these common methods, it is necessary to have some mathematical knowledge; however, they are explained in our book with figures and descriptions of the theory behind the software. In addition, we include numerical examples to instruct the reader on how existing popular software works. Thus, the use of mathematics is reduced and simplified. We argue this will be helpful for non-experts using established methods and will help avoid pitfalls and mistakes in application and interpretation. Finally, our book will assist the reader to understand and conceptualize the overall flow of brain imaging data analysis, particularly for statisticians and data-scientists unfamiliar with this area.

Software

R can be operated from the command line and can be reproduced by saving the command. This facilitates resumption of work and is useful for checking analyses results. Please refer to other documents because the basic usage of R is not included. The R package, mand, contains most functions used in this book. The details are described in this volume and can be downloaded from CRAN. https://cran.r-project.org/web/packages/mand/index.html

Summary

- Focuses on a multivariate approach, which is useful in not only brain imaging analysis but also (medical) big data analysis.
- Illustrates traditional (fundamental) statistical methods in brain imaging analysis that can be implemented by popular software utilizing an originally developed perspective to smoothly induce multivariate analysis.
- Demonstrates practical and easy-to-understand examples using the open-source, free statistical software R.
- Contributes useful techniques applicable for big data analysis in various biomedical fields.
- Provides an overview of a broad range of imaging analysis methods.

Acknowledgements

Dr. Takashi Yanagawa taught me a lot of things in my career as a researcher, such as research attitudes and research methods and has given me the opportunity to contribute my work to biostatistics, and I am very grateful to him. Dr. Young Truong guided our research on fMRI. I am very grateful to have met him and to have been able to work on this brain imaging study. He also made it possible for me to have a wonderful two years in Chapel Hill.

Dr. Kiyotaka Nemoto invited me to his annual lecture series as a lecturer, and much of the book's content is based on the lecture. He taught me what was needed in clinical research and we had a deep discussion about the lecture content. Dr. Fumio Yamashita has made a great contribution to our brain imaging analysis in the area of image processing. He is always discussing research topics in image analysis, and I've been able to share my research with him which is included in this book.

I would like to express my deepest gratitude to them for their advice on my education and research in brain imaging.

I would like to thank Vijay Primlani from CRC Press, who has worked with me over two years from an early book plan to production and reviewing the manuscript. I would like to thank Ryo Tajiri for carefully reading and reviewing the entire text. I think that the knowledge and skills gained with many collaborators have come together. We would like to thank all of our collaborators.

Finally, I would like to give special thanks to my wife, Machiko, for supporting me in my daily life, and to two sons, Saito and Junon, for their irreplaceable smiles.

Contents

Chapter 1

Introduction

Brain research is playing an important role worldwide by helping to characterize and find treatments for psychiatric disorders, such as dementia, schizophrenia, depression, and children's developmental disorders, that increasingly burden society. In particular, the early detection and treatment of Alzheimer's disease (AD) and Parkinson's disease are among the most prioritized research initiatives globally. Under these circumstances, statistical contribution is required to objectively evaluate the function and morphology of the brain.

The development of magnetic resonance imaging (MRI) has allowed for the study of brain structure and functions from images. Concurrent developments of statistical methodologies enable a more objective and quantitative evaluation of disease status that relies on questionnaires or cognitive tests and have helped advancing research by helping to reveal results that in turn have helped in the development of therapies. For example, dementia is a disease of the brain that deteriorates cognitive functions, such as memory and judgment, and quality of life. Referred to as "degenerative disease," the most common group of neurobiological diseases that causes a progressive death of neurons. One million Japanese patients reportedly suffer from AD, a type of degenerative disease. In recent years, several therapeutic drugs for Alzheimer's disease have been developed, making it possible to compensate for the decline in cognitive functions if the condition is detected early. In AD, brain atrophy begins from the nascent stages of the disease, and research is actively conducted to quantitatively characterize it using MRI and facilitate its early detection (Arbabshirani et al., 2017). The United States Alzheimer's Disease Neuroimaging Initiative is representative, of such efforts and has been extended to an international scale through the organization of the World-Wide Alzheimer's Disease Neuroimaging. Even at the international conference hosted by the Alzheimer's Association in July 2011, there was much interest in the elucidation of neuropathological progression to allow for early diagnosis and treatment.

In order to use brain imaging data obtained from clinical research, it is necessary to derive objective answers to the problem being studied by using statistics. An electronic brain image is displayed with color contrast, and numerical values assigned to each unit of the image (pixel). Therefore, statistical analysis can be performed using pixel values constituting the image as data; many analytical methods have been developed from this practice. Because of the simplicity of data interpretation and the limitations of executable software in the initial stages of brain imaging research, analyzing the data by applying a t-test, ANOVA and correlation analysis and generalizing it into a general linear model (GLM) for each pixel unit is common. In recent years, interest has focused on analyzing the patterns in the connections and relationships among neural structures since high dimensional data analytical methods, including applications such as artificial intelligence (AI), machine learning and regularization have been increasingly developed in the field of multivariate statistical analysis.

In this book, we'll focus on our own R package mand touching on the theory of analysis methods a few times. A package of R is an extension, which cannot be used just by installing R. It is necessary to install the package on R. Therefore, although R is often used in this book, we will leave the basic operations of R to other references for the sake of focus, but in this chapter, we will give an introduction to the package as well as practice of R operations. In the second half of this chapter, we will introduce how the image data are stored through R, how the brain image data and analysis results are displayed, and the features that may aid in the interpretation of the results.

There are many R packages that read, process, display, and analyze brain images. These are summarized in neuroconductor (`https://neuroconductor.org/`). It is also introduced in a well-organized manner in Polzehl and Tabelow (2019). The mand package depends on several other packages. For example, the msma package, which includes a related method of matrix decomposition among the multivariate analyses at the center of the book, has been developed by including methods that have been worked on by our group. Furthermore, the mand package depends on the caret package for predictive modeling and evaluation. These dependencies are incorporated simultaneously with the installation of the mand package.

There are many ways to measure brain images, as described further in the next section. Structural MRI (sMRI) and functional MRI (fMRI) are mainly treated as examples. However, the introduced analysis method is also applicable to data from fields other than brain imaging. The reason for this is that the subsequent method is almost the same after the data matrix for analysis is created from the

3-D brain images. It needs to transform from the original image to its analysis data matrix, which may be used as a result of the pre-processing of each commonly used software. We introduce the following four software programs for brain imaging analysis.

SPM (Statistical Parametric Mapping, http://www.fil.ion.ucl.ac.uk/spm/) is the most widely used software in MATLAB®, see Friston et al. (2007) and (Ashburner, 2012) for details. MATLAB which is required for SPM is commercial, but adding SPM is free. A number of toolboxes (extensions) are available, especially VBM8 and CAT12, which are useful for morphological analysis of brain images. Analysis of SPECT and PET images is also available. In this book, statistical methods used in SPM are introduced as common methods for brain imaging analysis. FSL (http://www.fmrib.ox.ac.uk/fsl/) is a Linux-based software developed by the analysis group at Oxford University, and is widely used in the same way as SPM. It is especially suitable for the analysis of images showing the degree of water diffusion in the brain. See Smith et al. (2004), Woolrich et al. (2009) and Jenkinson et al. (2012) for details. FreeSurfer (http://surfer.nmr.mgh.harvard.edu/) is often used to calculate brain volume from cortical thickness and anatomical segments (Fischl, 2012). Since many processes can be performed at once with simple commands, the usage in structural brain analysis is rapidly increasing. ANTs (advanced normalization tools, http://stnava.github.io/ANTs/) can perform advanced techniques for image transformation (Avants et al., 2009). Furthermore, various analysis methods have been incorporated in the preprocessing process, which are useful for brain imaging analysis. R packages have also been developed for each of these software programs, which allow users to operate them from R and visualize the analysis output.

R example

Preparation

This book is supported by the R statistical software. Elementary information about R such as installation, setup, and commands is omitted. The R homepage (https://www.r-project.org/) is a useful resource for beginners. The version of R used in this book is as follows.

```
R.version.string
```

```
## [1] "R version 3.5.3 (2019-03-11)"
```

The mand package provides several functions to implement the methods introduced in this book for manipulating brain imaging data. Once the R package is installed, additional functions become available. The first installation of the package should be performed from the Comprehensive R Archive Network (CRAN) using the following code.

```
if(!require("mand")) install.packages("mand")
```

Once the package is installed, this code is no longer required. The package is loaded with the following command line code.

```
library(mand)
```

This code must be executed every time to start R. Other packages are also available. Some depend on the mand package being pre-installed.

Template

One subject image as the template is available in the mand package. The code to load it is as follows.

```
data(template)
```

The image is compressed because of storage and computation time. The dimension is confirmed as follows.

```
dim(template)
```

```
## [1] 30 36 30
```

The image is plotted by the coat function.

```
coat(template)
```

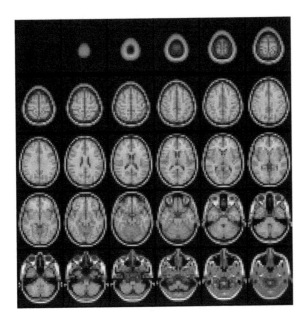

Other options with the plane argument (such as "axial," "coronal," "sagittal," and "all") are available. The default setting is "axial". If the argument is specified as `plane="all"`, three directional slices at a certain coordinate are plotted.

```
coat(x=template, plane="all")
```

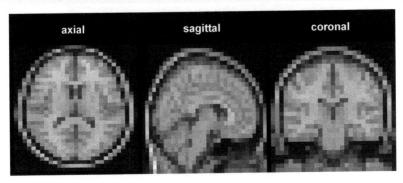

Image Data Matrix

The `imgdatamat` function reads image files saved in the nifti format and creates data matrices with subjects in rows and voxels in columns (this example does not work).

```
fnames1 = c("data1.nii", "data2.nii")
imgmat = imgdatamat(fnames1, simscale=1/4)
```

The first argument is file names with a length equaling the number of subjects (the number of rows in the resulting data matrix). The second argument simscale is the image resize scale. In this example, all the sizes (number of voxels) for the three directions were reduced to quarter size. The output is the list form where "S" is the data matrix, the "brainpos" is a binary image indicating the brain region, and the "imagedim" is the image dimension. The ROI (Region Of Interest) volume is computed in the "roi" if the ROI argument is TRUE.

```
imgmat = imgdatamat(fnames1, simscale=1/4, ROI=TRUE)
```

Overlay

The resulting map from the statistical analysis such as the "t" statistics map from the SPM is represented with overlapping on the template. For example, the mand package has the average difference representing Alzheimer's disease and healthy subjects with the array format.

The overlay is implemented by the coat function.

```
data(diffimg)
coat(template, diffimg)
```

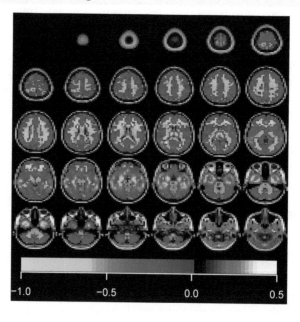

Atlas

The anatomical brain segmentation region is useful for the plot and its interpretation. For example, the Automated Anatomical Labeling (AAL) atlas is used. The data.frame has two columns ("ROIid" and "ROIname") format.

```
data(atlasdatasets)
atlasname = "aal"
atlasdataset = atlasdatasets[[atlasname]]
head(atlasdataset)
```

```
##   ROIid                      ROIname
## 0     0                   Background
## 1     1               Left Precentral
## 2     2              Right Precentral
## 3     3         Left Superior Frontal
## 4     4        Right Superior Frontal
## 5     5 Left Superior Frontal Orbital
```

It is also neccesary to prepare the array data.

```
data(atlas)
tmpatlas = atlas[[atlasname]]
dim(tmpatlas)
```

```
## [1] 30 36 30
```

It has ROIid as the element.

```
tmpatlas[11:15,11:15,10]
```

```
##      [,1] [,2] [,3] [,4] [,5]
## [1,]   56   56   56   56   56
## [2,]   98   98   98   56   56
## [3,]   98   98   96   96   96
## [4,]   98   98   96   96   94
## [5,]   98   96   96   96  110
```

The anatomical region can be plotted by the coat function with regionplot=TRUE.

```
coat(template, tmpatlas, regionplot=TRUE,
atlasdataset=atlasdataset, ROIids = c(1:2, 37:40),
regionlegend=TRUE)
```

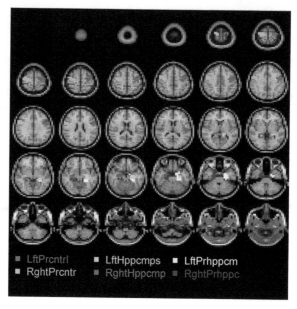

The resulting map can be converted into a summary of the anatomical regions.

```
atlastable(tmpatlas, diffimg, atlasdataset, ROIids = c(1:2,
37:40))
```

```
##      ROIid                   ROIname sizepct sumvalue    Min.
## 39     39   Left Parahippocampus   1.000  -20.387  -0.941
## 38     38       Right Hippocampus   1.000  -14.914  -0.912
## 37     37        Left Hippocampus   1.000  -17.864  -0.823
## 40     40  Right Parahippocampus   1.000  -20.822  -0.794
## 1       1           Left Precentral  0.643  -14.288  -0.312
## 2       2          Right Precentral  0.698  -16.033  -0.286
##          Mean Max.
## 39 -0.001    0
## 38  0.000    0
## 37 -0.001    0
## 40 -0.001    0
## 1   0.000    0
## 2   0.000    0
```

The output is the number of voxels in the `sizenum` column, the percentage of voxels in the `sizepct` column, and the minimum, mean, and maximum values of the region of the overlaying map. The order of the table row is in the larger absolute value off the minimum and maximum values.

The brain image corresponding to the region of interest can be extracted as follows. First, we create a mask image in which the hippocampal region is represented by 1 and others by 0. Then the product of the template and the mask image is taken for each voxel.

```
hipmask = (tmpatlas == 37) + (tmpatlas == 38)
template2 = template * hipmask
```

The images generated by these processes are plotted from left to right in one slice.

```
par(mfrow=c(1,3), mar=rep(1,4))
coat(template, pseq=11, paron=FALSE)
coat(hipmask, pseq=11, paron=FALSE)
coat(template2, pseq=11, paron=FALSE)
```

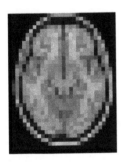  

The template image (left) and the mask image (middle) are multiplied voxel by voxel to obtain only the hippocampus region image (right).

The sum of the voxel values in the region is calculated as follows.

```
sum(template[which(hipmask==1, arr.ind = TRUE)])/1000
```

```
## [1] 185.9375
```

Such a value is calculated for each region and a dataset with ROI values is created.

Chapter 2

Brain Imaging Data

Modalities

In general, electronic images are matrices or arrays of numbers that correspond to spatial positions. In addition to MRI, brain imaging methods include positron emission tomography (PET), single photon emission computed tomography (SPECT), diffusion weighted imaging (DWI), and near-infrared spectroscopy (NIRS); each is used according to the purpose of the research. We focus on MRI in this book. This method irradiates the target object with radio waves in a strong static magnetic field, records signals emitted by resonance, and outputs them as images wherein each tissue of the brain is represented with varying signal intensities (data value) visualized by contrast (color intensity). Please refer to Filippi (2009) and Friston et al. (2007) for details.

By changing the precise method by which MRI is performed, images with different properties can be obtained. This book considers the following: (1) structural MRI (sMRI), which examines brain morphology (atrophy); and (2) functional MRI (fMRI), which indirectly obtains a blood oxygen level dependent (BOLD) signal from cerebral blood flow (hemoglobin in the blood) and records changes in cerebral blood flow as a correlation of underlying neural activity. Though other methods based on MRI are available, such as the analysis of metabolism and circulation with magnetic resonance spectroscopy, this book will not discuss them. Currently, not only normal tomographic images, but also images of various tissues and substances in the brain, such as imaging of nerve fibers based on diffusion anisotropy of water molecules in the brain (DWI), measurement of metabolites in the brain, and blood vessel images of the brain, have become an indispensable device for medical care.

Structural brain imaging data

The principles for obtaining the MRI signal are as follows. When a high-frequency magnetic field is applied to a tissue for a certain period, a phenomenon of nuclear spins called "excitation" is induced in the hydrogen atoms that constitute water molecules and fat in the tissue. After that, a phenomenon called "relaxation", which is the opposite of excitation, occurs when the electromagnetic waves are stopped. The magnetic resonance signal is obtained by measuring the magnetic field emitted at this time using a receiving coil. The time required for relaxation (relaxation time) varies from tissue to tissue, so that the intensity of the magnetic resonance signal at a certain time (echo time: TE) after excitation depends on the tissues in the brain, such as white matter, gray matter, and ventricles (=cerebrospinal fluid). If we assign a grayscale to each voxel intensity, we can obtain an image of the brain structure. Since the magnetic resonance signal is obtained by exposing a radiofrequency magnetic field of the same frequency as the nuclear spins, it is possible to selectively excite a three-dimensional slice of the brain by tilting the magnetic field at an angle to the coil. Then, a 3D brain image is obtained by using the Fourier transform to obtain the location information and reconstructing it.

Figure 2.1(A) shows that when the human brain is represented by an image, it is partitioned in units called voxels (pixels in two dimensions) which are a product of horizontal, vertical, and slice. The resolution is determined by the size of the voxel. In case the size of the voxel is 1.5 mm^3 - common in research - the number of voxels of the structural image after having been fitted to the reference image is 121 (horizontal) × 145 (vertical) × 121 (slice) = 2,122,945 voxels. Differences in brain structures such as gray matter and white matter can be ascertained from

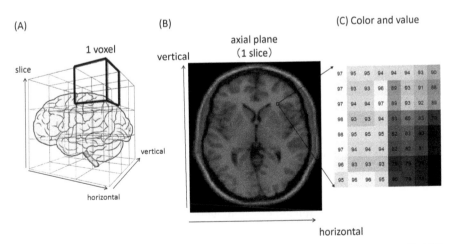

Figure 2.1: (A) 3 (horizontal) × 3 (vertical) × 3 (slices) = 9 voxels of 3D brain image. (B) MRI morphological image in axial plane. (C) 8 × 8 = 64-pixel two-dimensional image (numerical values are pixel values) with the square part of (B) enlarged.

the data values obtained for each voxel. Though the data value stored in each voxel is referred to by various means, including brightness and pixel value, this review calls it the voxel value. Figures 2.1(B) shows a morphological image of the human brain obtained with MRI as most frequently presented: axial plane, two-dimensional representation.

Functional brain imaging data

fMRI has been widely applied in the fields of neurosurgery, neurology, internal medicine, cognitive psychology, child development, rehabilitation medicine, and reproduction science among others. One of the fields using fMRI is Parkinson's disease research. Parkinson's disease is caused by dopamine deficiency in the brain, causing symptoms such as trembling and difficulty in movement. Mergers with dementia also occur. Many books and reviews on statistical analysis of fMRI data have been published. Friston et al. (2007) is a representative example that also discusses analysis using SPM. In addition, Lazar (2010), Poldrack et al. (2011), Ashby (2019), Polzehl and Tabelow (2019) treats the statistical analysis of fMRI data. Lindquist (2008) and Chen and Glover (2015) are review articles.

fMRI measures changes in blood flow in the brain. The increase in hemoglobin to compensate for that consumed in the blood (oxidized) by brain activity is captured as a signal. This is called the blood oxygen level dependent (BOLD) effect; fMRI measures the change in the MR signal effected by BOLD to identify active parts of the brain. fMRI is spatiotemporal (4D) data: 3D brain images are measured over time. One voxel is 3 mm^3, and 64 × 64 × 49 (voxel) × 100 (time points) is a standard setting. The example in Figure 2.2 is called block design: the subject is periodically given a task, and the attendant brain activity is recorded. The voxel value time series related to the task correlates with the problem time series taking binary values of on and off. Sen et al. (2010) and Lewis et al. (2011) examined brain functions when research participants engaged in finger tapping with different types and speeds. There are other event-related designs proposed by Buckner et al. (1996).

Sen et al. (2010) and Lewis et al. (2011) uses the mixed effect model as described in Section 3 of a Chapter 3 clause to investigate the effects of the thalamic-striatal-cortical pathway and the cerebellar-thalamocortical pathway in Parkinson's disease. In Taniwaki et al. (2007), network analysis of the basal ganglia motion path related to spontaneous movement was performed using the structural equation modeling introduced in Chapter 5.

Figure 2.2 is called task-related research, which is conducted with various stimuli and tasks, and conversely, resting state research, which is conducted in a resting state without any stimuli. In recent years, network analysis has been widely used to evaluate the functional connectivity between resting sites. Please refer

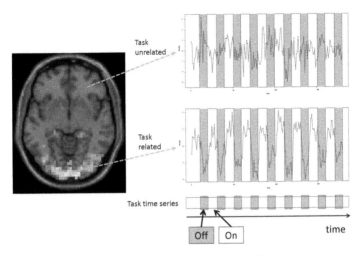

Figure 2.2: Voxel value time series in fMRI data.

to Chapter 5 for the analysis. The medial prefrontal cortex and the anterior cuneiform and posterior cingulate gyrus are a network of brain regions called the default mode network (DMN), and they are more active when the subject is awake but not doing anything (Raichle et al., 2001). This neural activity consumes 60–80% of the energy in the brain. However, on the other hand, the energy expenditure of neural activity is only 0.5–1.0% in the brain while performing some kind of task. The DMN has been studied in various psychiatric disorders as it is believed to be involved in internal thought processes and the control of external stimuli (Joo et al., 2016; Liu et al., 2017). In addition to the DMN, several other networks showing activity at rest have been identified and resting-state fMRI is thought to contain a lot of information about individual brain function. Tavor et al. (2016) showed that an individual's functional connectivity pattern of resting fMRI can predict the pattern of fMRI activity while performing a task.

Multivoxel pattern analysis (MVPA) has been actively used to examine the brain activity patterns using multivariate analysis and machine learning, where brain activity patterns are used as explanatory variables and stimuli and tasks are used as objective variables (Mahmoudi et al., 2012). We use MVPA to analyze the differences in the spatial activity patterns of multiple voxels by considering the stimuli and tasks as encoded. Then, we evaluate whether the pattern of brain activity contains identifiable information and decoding. This kind of analysis is called brain information decoding (Haxby et al., 2014).

In addition to fMRI, magnetoencephalography (MEG), which can record detailed electrical activity of the brain of the millisecond (ms) order, has been used to evaluate brain function (Gross, 2019). While fMRI measures metabolic changes associated with brain activity, MEG can measure brain activity in real time and is suitable for time series and network analysis. MEG is a more direct measure of neuronal activity than fMRI. However, while fMRI can identify which part of the brain caused the change in activity, MEG provides only an estimate. Therefore, a combination of both is useful for analyzing brain function. A review of methods specific to AD research is provided by Mandal et al. (2018).

Format

There are several formats for storing images measured by MRI as electronic files on a computer. The digital imaging and communication in medicine (digital imaging and communications in medicine: DICOM) format is common for medical images. By saving images captured by different devices in this format, the user can see images without being conscious of the difference in the devices used. "Analyze" is a format consisting of a data file (.img) that records pixel information and a header file (.hdr) that records meta information of the origin information as well as the size of the image. However, in the Analyze format, it was not possible to record the relation between the voxel address and the spatial information of the MRI machine. This limitation is overcome in the neuroimaging informatics technology initiative (NIfTI) format, an extension of the Analyze format that has become mainstream in recent years. As with Analyze, NIfTI can store data files and header files separately, and these two files can be saved together (.nii).

Several statistical methods can be applied when image data is converted for statistical analysis. As an example, the pixel image of 2×2 for four subjects and its pixel values are shown in Figure 2.3(I). The higher the pixel value, the brighter

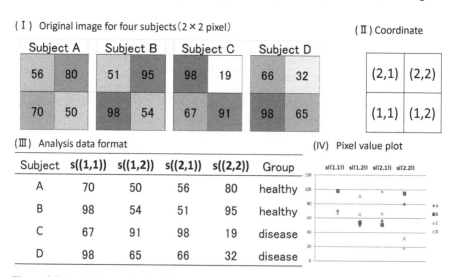

Figure 2.3: 2×2 images for 4 subjects. The numbers represent pixel values. Format for statistical analysis of image data (III) and its plot (IV).

the color; and the lower the pixel value the darker the color. The image is thus represented by color contrast.

Let v be the vector representing the coordinate with $v_1 = (1, 1)$, $v_2 = (1, 2)$, $v_3 = (2, 1)$, $v_4 = (2, 2)$. The pixel data of the α-th subject is represented as $s_\alpha = (s_\alpha(v_1), s_\alpha(v_2), s_\alpha(v_3), s_\alpha(v_4))^\top$. If pixel values for one pixel are regarded as one variable and are organized column by column, a data matrix like that in Figure 2.3(III) can be obtained.

Furthermore, as in Figure 2.3(III), information on the participants in the healthy or disease groups can be incorporated into the column. When plotting Figure 2.3(III), it looks like Figure 2.3(IV), and in $v_4 = (2, 2)$, there is a difference in pixel values between groups. Thus, by conducting an average value test on the pixel values for each coordinate, it is possible to investigate differences between the healthy and diseased groups. Details are explained in Chapter 3. Alternatively, by making the health status of the participants the objective variable, disease can be predicted as a discrimination problem. In this case, the explanatory variables are high dimensional data consisting of millions of voxels. This analysis is summarized in Section 4 of Chapter 4.

Preprocess

Images are not used for analysis as obtained; they require preprocessing that helps to facilitate analysis, satisfies the assumption of the chosen statistical method, and reduces noise. Preprocess is roughly divided into the followings: (1) segmentation, (2) anatomical standardization (spatial normalization), and (3) smoothing. In others, realignment and slice timing correction are specified in fMRI data analysis. Details of this process are summarized in Friston et al. (2007).

Segmentation of sMRI images is mainly preformed during preprocess: images are divided into gray matter, white matter, and cerebrospinal fluid based on voxel values. This is also called brain tissue classification (tissue classification). Since it is unknown to which category each voxel belongs to, segmentation is a problem of clustering in the field of statistics. A method based on a normal mixture distribution model is common, and the Bayes method is used as advance information obtained in the past and accuracy is improved by combination with other preprocessing methods (Ashburner and Friston, 2005a). Recently, Feng et al. (2012) has been devised to improve the accuracy of segmentation by using the Hidden Markov Model. The term segmentation is also used to compose anatomical divisions (brain maps, brain atlas) from brain images; this process is summarized by Cabezas et al. (2011). In addition, the statistical evaluation index in segmentation is summarized in Popovic et al. (2007). Certain considerations must be taken into account prior to segmentation because images are distorted due to the nonuniformity of the magnetic field that adversely affects the voxel-value based segmentation. Correction for this is called distortion correction. A method for achieving distortion correction based on several nonlinear transformation equations has been proposed. GradunWarp by Jovicich et al. (2006) is a method that recruits a spherical harmonics function and is used in the United States Alzheimer's Disease Neuroimaging Initiative. Alternative methods include those using thin plate spline (Bookstein, 1989, Dryden and Mardia, 1998), and Ashburner and Friston (2005b), which are integrated ways of accomplishing alignment with the standard brain.

Anatomical standardization aligns voxels to standard brain images. This can be regarded spatially as standardization (position, size, shape matching); voxels with the same coordinates can be regarded as corresponding to the same part of the brain, even for different subjects. This allows for the comparison among subjects and is useful for the interpretation of analysis results. Concerning anatomical standardization for sMRI analysis, many methods based on nonlinear transformation have been proposed. If the required standardization is simple, affine transformation and discrete cosine transformation can be used.

Shen and Davatzikos (2002) proposed hierarchical attribute matching mechanism for elastic registration (HAMMER) based on elastic transformation, while the differential-inomorphism based SyN (symmetric diffeomorphic registration, Avants et al., 2008; Avants et al., 2011) is used in the software ANTS. Diffeomorphic anatomical registration using exponentiated lie (DARTEL) algebra, Ashburner, 2007) is normally equipped in SPM software, version 8 or beyond. For details on these methods, please refer to each of the aforementioned documents. In addition to these, 14 standardization methods are compared and examined in Klein et al. (2009), and it is reported that SyN and DARTEL generally have good accuracy. In addition, when the brain volume is calculated from the image and used for analysis, it is necessary to correct the voxel value so as to maintain the post-standardization volume. This correction method is called modulation.

Transformation

The obtained image is not directly used for analysis; instead, some preprocessing is performed. This process has various benefits such as to help with interpreting the analysis results, to meet the assumptions of the application method and to reduce noise. To interpret the results, especially when comparing pixel values between subjects as in the previous section pixels must represent the same part. Here, it is being used as a basic method for alignment. This section describes the affine transformation for changing the position (called geometric transformation) and the interpolation method necessary for changing the resolution.

The affine transformation performs linear transformation of coordinates and balanced movement at the same time. The transformation from coordinates (x, y) to (x', y') is given by

$$\begin{pmatrix} x' \\ y' \end{pmatrix} = T \begin{pmatrix} x \\ y \end{pmatrix} + s, \text{ where } T = \begin{pmatrix} a & c \\ b & d \end{pmatrix}, s = \begin{pmatrix} e \\ f \end{pmatrix}$$

The matrix T represents a linear transformation, and the vector s represents a slide, for example, see the middle portion of Figure 2.4. Consider the transformation of a rectangle with its vertices at coordinates A $(-1, -1)$, B $(-1, 1)$, C $(1, -1)$, and D $(1, 1)$.

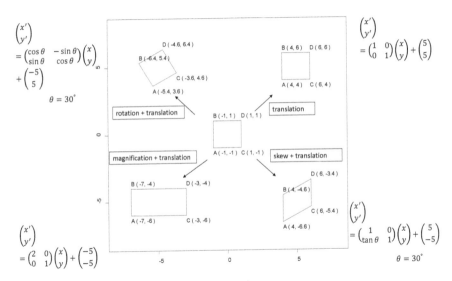

Figure 2.4: Affine transformation.

Since $a = b = c = d = 0, e \neq 0, f \neq 0$ represents a translation if $a = b = c = d = 0$ and $e = f = 5$; this resembles the upper right portion of Figure 2.4. When $a = \cos\theta, b = \sin\theta, c = -\sin\theta$, and $d = \cos\theta$, the rotation angle is θ. When $\theta = \pi/6$, $e = -5$, and $f = 5$, rotation and translation are performed as shown in the upper left portion of Figure 2.4. $a \neq 0, b = c = 0, d \neq 0$. When $a = 2, = c = 0, d = 1, e = -5$, and $f = -5$, the horizontal axis expands and translates as shown in the lower left of Figure 2.4. When $a = 1, b = \tan\theta, c = 0$, and $d = 1$, it is skewed (parallelogram). When $a = 1, b = \tan\pi/6, c = 0, d = 1, e = 5$, and $f = -5$, it is parallel to the skew, as shown in the lower right of Figure 2.4.

Affine transformation

The affine transformation of the figure is explained with R code. Basically, it is a matrix operation. At the beginning a matrix whose elements are the coordinates to be transformed is prepared.

We begin by preparing a rectangle in a two-dimensional plane as an example. The coordinates of each vertex are given as a matrix.

```
X = matrix(c(
-1, -1, 1, 1,
-1, 1, -1, 1
),ncol=4, byrow=TRUE)
```

A square is plotted with the coordinates in the matrix as vertices.

```
par(mfrow=c(1,1), cex=0.4)
plot(NULL, xlim=c(-8, 8), ylim=c(-8, 8), xlab="x", ylab="y")

rect(xleft=min(X[1,]), ybottom=min(X[2,]),
xright=max(X[1,]), ytop=max(X[2,]))

vs = sapply(1:4, function(x)
paste("(", paste(X[,x], collapse=", "), ")"))
text(X[1,], X[2,], paste(LETTERS[1:4], vs), pos=c(1,3,1,3))
```

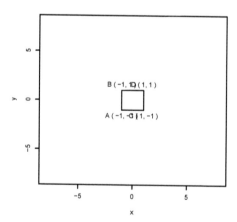

For the coordinate matrix, we perform parallel movement as follows. The rectangle before and after the move is displayed with the coordinates.

```
Y = X + c(5,5)
```

```
par(mfrow=c(1,1), cex=0.4)
plot(NULL, xlim=c(-8, 8), ylim=c(-8, 8), xlab="x", ylab="y")
rect(xleft=min(X[1,]), ybottom=min(X[2,]),
xright=max(X[1,]), ytop=max(X[2,]))
rect(xleft=min(Y[1,]), ybottom=min(Y[2,]),
xright=max(Y[1,]), ytop=max(Y[2,]))
vs = sapply(1:4, function(x)
paste("(", paste(Y[,x], collapse=", "), ")"))
text(Y[1,], Y[2,], paste(LETTERS[1:4], vs), pos=c(1,3,1,3))
```

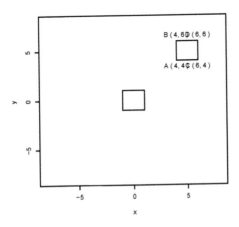

Next, we prepare the matrix T and perform enlargement and further parallel movement.

```
T = matrix(c(
2, 0,
0, 1
),ncol=2,byrow=TRUE)
s = c(-5,-5)
Y = T %*% X + s
```

```
par(mfrow=c(1,1), cex=0.4)
plot(NULL, xlim=c(-8, 8), ylim=c(-8, 8), xlab="x", ylab="y")
rect(xleft=min(X[1,]), ybottom=min(X[2,]),
xright=max(X[1,]), ytop=max(X[2,]))
rect(xleft=min(Y[1,]), ybottom=min(Y[2,]),
xright=max(Y[1,]), ytop=max(Y[2,]))
vs = sapply(1:4, function(x)
paste("(", paste(Y[,x], collapse=", "), ")"))
text(Y[1,], Y[2,], paste(LETTERS[1:4], vs), pos=c(1,3,1,3))
```

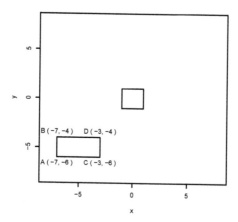

Next, we prepare the rotation matrix T and perform the rotation and parallel movement.

```
theta = pi/6
T = matrix(c(
cos(theta), -sin(theta),
sin(theta), cos(theta)
),ncol=2,byrow=TRUE)
s = c(-5, 5)
Y = T %*% X + s

par(mfrow=c(1,1), cex=0.4)
plot(NULL, xlim=c(-8, 8), ylim=c(-8, 8), xlab="x", ylab="y")
rect(xleft=min(X[1,]), ybottom=min(X[2,]),
xright=max(X[1,]), ytop=max(X[2,]))
lines(Y[1, c(1, 2)], Y[2, c(1, 2)])
lines(Y[1, c(1, 3)], Y[2, c(1, 3)])
lines(Y[1, c(2, 4)], Y[2, c(2, 4)])
lines(Y[1, c(3, 4)], Y[2, c(3, 4)])
vs = sapply(1:4, function(x)
paste("(", paste(round(Y[,x],1), collapse=", "), ")"))
text(Y[1,], Y[2,], paste(LETTERS[1:4], vs), pos=c(1,3,1,3))
```

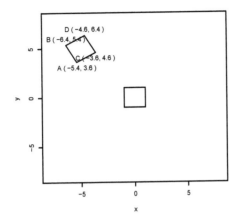

Finally, another matrix T is prepared to perform the distortions and parallel moves.

```
T = matrix(c(
1, 0,
tan(pi/6), 1
),ncol=2,byrow=TRUE)
s = c(5, -5)
Y = T %*% X + s

par(mfrow=c(1,1), cex=0.4)
plot(NULL, xlim=c(-8, 8), ylim=c(-8, 8), xlab="x", ylab="y")
rect(xleft=min(X[1,]), ybottom=min(X[2,]),
xright=max(X[1,]), ytop=max(X[2,]))
lines(Y[1, c(1, 2)], Y[2, c(1, 2)])
lines(Y[1, c(1, 3)], Y[2, c(1, 3)])
lines(Y[1, c(2, 4)], Y[2, c(2, 4)])
lines(Y[1, c(3, 4)], Y[2, c(3, 4)])
vs = sapply(1:4, function(x)
paste("(", paste(round(Y[,x],1), collapse=", "), ")"))
text(Y[1,], Y[2,], paste(LETTERS[1:4], vs), pos=c(1,3,1,3))
```

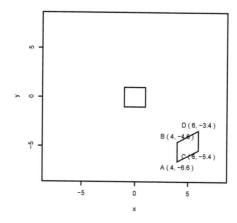

Resolution

Next, resolution conversion is explained. Assume that the pixel value of an image with a resolution of 2 × 2 is as shown in Figure 2.5 (A). This is converted to a 6 × 6 resolution as shown in Figure 2.5(B). The four pixel values in Figure 2.5(A) must then be spaced as shown in Figure 2.5(B), and blank pixels will be interpolated. Here, two interpolation methods are introduced. Figure 2.5(C) is the nearest neighbor method and interpolates with the pixel value of the nearest address. Figure 2.5(D) is a linear interpolation method. In Figure 2.5(B), the pixels

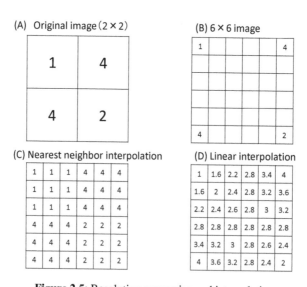

Figure 2.5: Resolution conversion and interpolation.

with values are interpolated with pixel values equally spaced between the maximum and minimum. Other methods include quadratic interpolation and spline interpolation. Conversely, there is a conversion from the high resolution of Figure 2.5(C) and (D) to the low resolution of Figure 2.5(A). One common method is to average the merged pixel values. Decreasing such resolutions also reduces the dimension being analyzed.

There are many references related to image processing that provide details on the geometric transformations and interpolation introduced here, for example, refer to the document Oliveira and Tavares (2014).

Resize

We introduce an R code to change the resolution. The sizechange function of the mand package is used. The image data and magnification are the inputs. Here, the same magnification is specified for length, width, and height, but it can be different. The first case is when the magnification is set to 4.

```
simscale1 = 4
img1r = sizechange(template, simscale=simscale1)
dim(img1r)
```

```
## [1] 120 144 120
```

```
coat(img1r, plane="all")
```

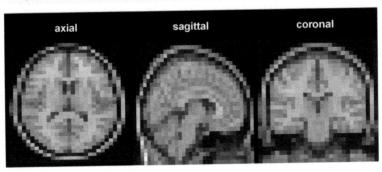

The next case is when the magnification is set to 1/2.

```
simscale1 = 1/2
img1r = sizechange(template, simscale=simscale1)
dim(img1r)
```

```
## [1] 15 18 15
```

```
coat(img1r, plane="all")
```

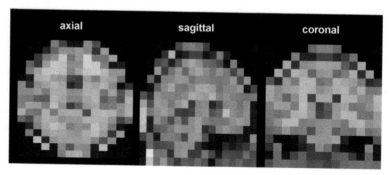

When the resolution is reduced in this way, the image becomes rough and information loss occurs. In the analysis, it is a lower dimension and reduces the computational cost, but the results may not tell us much about the details.

Smoothing

Smoothing is performed as the final stage of pretreatment to suppress excessive fluctuation of voxel values within an individual image and to apply Gaussian random field theory. In SPM—described later–smoothing is performed by convolution using a Gaussian function.

The Gaussian function is defined by two parameters as given in the following equation.

$$f(x) = \frac{1}{\sqrt{2\pi}\sigma} \exp\left[-\frac{(x-\mu)^2}{2\sigma^2}\right]$$

In statistics, it is called the density function of the normal distribution with mean μ and standard deviation σ, but in brain image analysis, the central coordinate is μ and the spread is defined by the full width at half maximum (FWHM). FWHM is expressed by the standard deviation σ. This can be obtained as follows. As the maximum is $f(\mu) = 1/(\sqrt{2\pi}\sigma)$, we find the value of "$x$" that is half of the maximum.

$$\frac{1}{\sqrt{2\pi}\sigma} \exp\left[-\frac{(x-\mu)^2}{2\sigma^2}\right] = \frac{1}{2}\frac{1}{\sqrt{2\pi}\sigma}$$

After that, solve for x.

$$\exp\left[-\frac{(x-\mu)^2}{2\sigma^2}\right] = \frac{1}{2} \implies (x-\mu)^2 = 2\log(2)\sigma^2$$

$$\implies x = \mu \pm \sqrt{2\log(2)}\sigma$$

The difference between these two solutions is the FWHM

$$FWHM = 2\sqrt{2\log(2)}\sigma \approx 2.35482\sigma$$

where $\sqrt{2\log(2)} = 1.17741$ is used in the computation.

1D Gauss function

The density function of a standard normal distribution with a mean of 0 and a standard deviation of 1 is plotted. The FWHM is then calculated as 1.17741 (in the R code, input and output are performed simultaneously by enclosing in parentheses) and is indicated by a dotted line in the figure.

```
(a = sqrt(2*log(2)))
```

```
## [1] 1.17741
plot(dnorm, -4, 4, xaxt = "n", ylab="f(x)")
axis(1, -4:4, -4:4)
lines(c(0,0), c(-1,dnorm(0)), lty=2, col=1)
lines(c(-a,a), c(dnorm(0)/2,dnorm(0)/2), lty=2, col=1)
lines(c(a,a), c(-1,dnorm(a)), lty=2, col=2)
lines(-c(a,a), c(-1,dnorm(a)), lty=2, col=2)
```

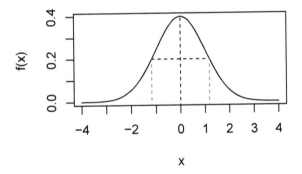

Preparation

The example data is generated from a one-dimensional random field by the following code.

```
set.seed(1)
n = 20
x = 1:n
y = abs(rnorm(n)); names(y) = x
barplot(y)
```

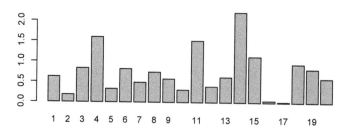

In order to implement examples below, two functions are defined here. The first is the gauss function with the coordinates x and the center coordinate x1.

```
g1 = function(x, x1, FWHM){
sigma = FWHM / 2*sqrt(2*log(2))
d1 = dnorm(x, x1, sigma)
names(d1) = x
d1
}
```

The second is the smoothing function using the gauss function.

```
gsmooth = function(x, y, FWHM){
sigma = FWHM / 2*sqrt(2*log(2))
sy = sapply(x, function(x1)
weighted.mean(y,
dnorm(x, x1, sigma)/sum(dnorm(x, x1, sigma))) )
names(sy) = x; sy; }
```

Smoothing flow

The original data is in the first row of the plot with data (signal) in the vertical line and the coordinates in the horizontal line, the gauss function with the 5th coordinate as the center in the second row, the convoluted data is in the third row, and summations are in the fourth row.

```
x1 = 5
col1 = rep(1, n); col1[x1] = 2
par(mfrow=c(4,1), mar=c(3,4,1,4))
barplot(y, col=col1)
barplot(g1(x,x1,2), col=col1, ylim=c(0, 0.4))
barplot(y*g1(x,x1,2), col=col1)
```

```
sy = rep(0, n); sy[x1] = sum(y*g1(x,x1,2)); names(sy) = x
barplot(sy, col=col1, ylim=c(0,2))
```

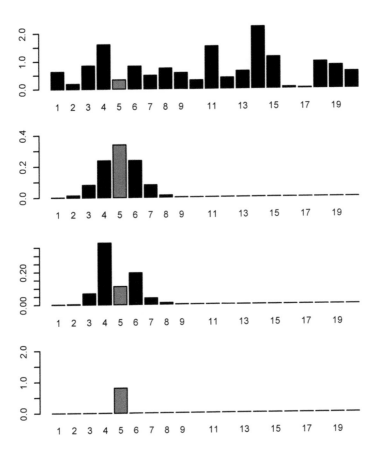

The above flow is applied to all coordinates as the center and the smoothed data with FWHM=2 was obtained in the second row panel.

```
col2 = rep(1, n)
par(mfrow=c(2,1))
barplot(y, col=col2, main="Original")
f1 = 2
sy = gsmooth(x, y, f1)
barplot(sy, main=paste("FWHM =", f1), col=col2, ylim=c(0,2))
```

Original

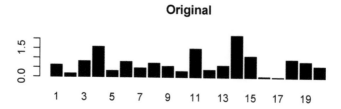

FWHM = 2

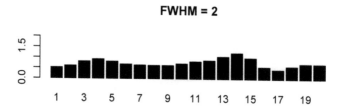

FWHM and smoothed data

The Gauss functions with different FWHM values are displayed on the left side. The different FWHMs were examined to smooth the original data and obtained the smoothed data as follows.

```
f1s = c(2, 4, 8)
layout(cbind(c(8,1,2,3), c(4:7)))
par(mar=c(3,4,2,4))
for(f1 in f1s){
barplot(g1(x,8,f1), main=paste("FWHM =", f1), ylim=c(0,0.4))
}
barplot(y, main="Original")
for(f1 in f1s){
sy = gsmooth(x, y, f1)
barplot(sy, main=paste("FWHM =", f1), ylim=c(0,2))
}
```

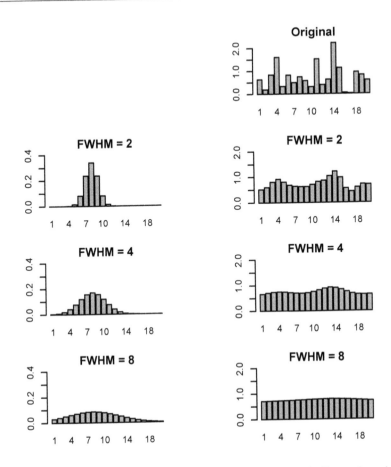

The larger FWHM values yielded more smooth signals (similar values in the neighborhood).

Other methods

Realignment is often used to achieve consistency among a series of fMRI images; this mainly entails head-motion and slice-timing corrections. Concerning the former, individuals receiving fMRI are instructed not to move their heads or bodies during imaging, but minute shifts are inevitable. In statistical analysis, the brain position of each subject under measurement is assumed to be constant. It is therefore necessary to correct any displacements across images. As a correction method, a rigid body transformation having six parameters consisting of parallel movement (x, y, z) and rotation (pitch, roll, yaw) is used. The images at each point in time are combined with the reference image (generally the image obtained at the first time point). To optimize parameters, a least-squares method, correlation coefficient, and mutual information have been employed. Mutual information seems to be used in major optimizations.

In slice corrections, fMRI images are taken over time, but it takes several seconds to take images at one time point in a given slice direction. Since it is assumed that the images obtained at one point are taken at the same time, it is necessary to correct the time difference among the images. As a correction method, temporal interpolation (linear interpolation, spline interpolation, sinusoidal curve interpolation) is performed by regarding pixel values of voxels as time series data. A method of considering time difference in statistical analysis without slice correction in preprocess has also been proposed (Friston et al., 1998).

Chapter 3

Common Statistical Approach

Structural brain imaging analysis

The sMRI is used to assess brain atrophy in dementia research. Voxel based morphometry (VBM, Ashburner and Friston, 2000), which has been employed to compare the local volume of gray matter between patients with AD and healthy controls, is used most frequently in clinical research as an analytical method. There are other means of assessing brain disease, such as surface-based morphometry (SBM) measuring cortical thickness, and using diffusion tensor images with tract-based spatial statistics (TBSS). The common statistical analysis method is a GLM, followed by multiple correction and random field theory (RFT) for making inferences. We will introduce these in the following sections. These can be executed by any of the softwares described in Chapter 1.

General linear model

As represented by VBM, a method that applies a GLM for each voxel and displays statistics calculated from the estimated parameters is often used. The objective variable is the voxel value in the voxel k, an n number of subjects are assumed, and the objective variable vector is defined as $\boldsymbol{Y}_k = (Y_{k1}, Y_{k2}, \ldots, Y_{kn})^\top$. $k = 1, 2, \ldots, V$, and V is the number of voxels. On the other hand, consider the following model.

$$\boldsymbol{Y}_k = \boldsymbol{X}\boldsymbol{\beta}_k + \boldsymbol{\varepsilon}_k, \quad (k = 1, 2, \ldots, V) \tag{3.1}$$

Here, X is the $n \times p$ design matrix, and, in many cases, the columns of X are grouped (a disease group and a healthy group as in Figure 3.1), and covariates. β_k is a p dimensional regression parameter, $\varepsilon_k \sim N(\mathbf{0}, \sigma_k^2 I)$ is an error term. If the design matrix X is properly given, the regression parameters are estimated by the least squares method and are obtained as $\hat{\beta}_k$ as follows.

$$\hat{\beta}_k = (X^\top X)^{-1} X^\top Y_k$$

The unbiased estimator for σ_k^2 is given as $\hat{\sigma}_k^2 = e_k^\top e_k / (n - (p+1))$ where e_k is the residual vector, $e_k = Y_k - X\hat{\beta}_k$. The estimation of such a comparison between groups is performed using contrast. For the contrast vector c, the test statistic for the null hypothesis $H_0 : c\beta_k = 0$ is given by

$$T_k = \frac{c^\top \hat{\beta}_k}{\sqrt{c^\top \text{Var}[\hat{\beta}_k] c}} \sim t_{n-p}, \quad (k = 1, 2, \ldots, V) \tag{3.2}$$

where $\text{Var}[\hat{\beta}_k]$ is the variance covariance matrix of $\hat{\beta}_k$. These t statistics are obtained for each voxel. By plotting these t-statistics on a brain image after establishing a certain threshold value, a brain region with a significant difference is visualized. Further, expansion of this model is considered, repeated measured data analysis has also been carried out. Ziegler et al. (2012) summarizes modeling of age-related changes in the brain by nonlinear regression analysis that extends the GLM.

Application example of GLM

An example of GLM using SPM version 12 (SPM12) is provided. Image data are from a database published by the Open Access Series of Imaging Studies (OASIS-1, http://www.oasis-brains.org/) by the Alzheimer's Disease Research Group at the University of Washington. Subjects were clinically diagnosed by the clinical dementia scale (CDR) international assessment method (CDR; normal (= 0), doubt (= 0.5), mild (= 1), moderate (= 2), high (= 3)). Subjects with CDR$\geq$1 were suspected of Alzheimer's disease (AD) and those with CDR = 0 were considered normal cognitive (NC). Fourty-four subjects (22 AD patients and 22 NC subjects) were selected with good quality of gray matter T1-weighted image, matching age, gender, brain volume, and educational history. A comparison between the AD and NC groups was performed by GLM. Anatomical standardization and smoothing were performed as preprocessing and images were analyzed using CAT12 (computational anatomy toolbox), an SPM extension toolbox. Since the voxel size was 3 mm cubed, the number of voxels after processing was $91 \times 109 \times 91 = 902,629$. The smoothing parameter, FWHM, was set to 8 mm.

Figure 3.1 shows the result of fitting GLM using SPM. On the right side of Figure 3.1, a graphical representation of the design matrix is output. Assume that the subscript i for the subject is rearranged and allocated—i.e., let $i = 1, 2, \ldots, 22$ be healthy people and $i = 23, 30, \ldots, 44$ be patients with AD. At this time, the design matrix $\boldsymbol{X}$ in the model expression (3.1) is 44×3, the first column is $(\boldsymbol{1}_{22}^{\top}, \boldsymbol{0}_{22}^{\top})^{\top}$, the second column is $(\boldsymbol{0}_{22}^{\top}, \boldsymbol{1}_{22}^{\top})^{\top}$, and the third column is the whole brain volume value, which is a covariate, where $\boldsymbol{1}_a$ is a vector whose elements of length a are all 1, and $\boldsymbol{0}_a$ has elements of length a all being 0 vectors. Inter-group comparison after adjustment for the whole brain volume is a contrast vector $\boldsymbol{c} = (1, -1, 0)^{\top}$ in the test statistic of the expression (3.2).

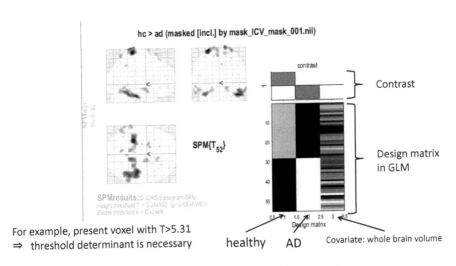

Figure 3.1: GLM application example (SPM output).

The test statistic for each voxel is plotted on the brain image template as shown on the left side of the Figure 3.1. In this case, the threshold of the test statistic is 5.31—i.e., only voxels whose test statistic is larger than 5.31 are filled with the color density corresponding to the magnitude of the test statistic. The threshold of 5.31 corresponds to FWER $= 0.05$ in the next section and is set to the default of FWER in SPM. The anatomical brain region is specified from the coordinates of the black part of the brain image, and the result is analyzed. Therefore, depending on the threshold of the test statistic, the result differs. It is therefore necessary to objectively determine the threshold. This is described in the next section.

Mixed effects model

As with sMRI, the most widely used analytical method is GLM. However, in the case of fMRI, a two-step method of fitting a model within an individual and comparing between individuals is used (Mumford and Nichols, 2006; Monti, 2011). Let $Y_i = (Y_{i1}, Y_{i2}, \ldots, Y_{iT})^\top$ be the BOLD signal of a certain voxel in subject i (subscripts on voxels are omitted). In this case, T is the measurement point number. Then consider the following GLM.

$$
\begin{aligned}
Y_i &= X\beta_i + \varepsilon_i, \quad (i = 1, 2, \ldots, n) \qquad &(3.3)\\
\beta &= X_G\beta_G + \varepsilon_G \qquad &(3.4)
\end{aligned}
$$

However, for the expression (3.3), X is the design matrix of $T \times 1$, and the columns correspond to those of the assignment β_i which is a regression parameter, $\varepsilon_i \sim N(0, \sigma_i^2\Sigma)$ is the T dimensional error term vector. $\beta = (\beta_1, \beta_2, \ldots, \beta_n)^\top$ for the expression (3.4), X_G is the design matrix of $n \times p$ (such as subject group and covariates), β_G is the p dimensional regression parameter, $\varepsilon_G \sim N(0, \sigma_G^2 I)$ is the n dimensional term vector. This is a so-called mixed effects model, where the expressions (3.3) and (3.4) are referred to as the 1st and 2nd levels, respectively, according to the SPM designation.

At the 2nd level, the same theory as that applied in the case of sMRI, like the cluster level and RFT described in Section 3 of this chapter, can be applied; however, at the 1st level, analysis of fMRI varies from that of sMRI in the design matrix and the error term that it employs. With respect to the design matrix, it is necessary to take into consideration that the BOLD signal of the objective variable does not simply describe the neural response, and a delay occurs. This phenomenon is represented by a hemodynamic response function (HRF, Worsley and Taylor, 2006). HRF depends only on TR (repetition time; imaging time of whole brain, usually 2, 3 seconds), which is one of imaging parameters. In this way, each column of the design matrix is a convolution integral of a column representing task and HRF: if $v(t)$ is the task time series and $h(t)$ is HRF, the design matrix column $x(t)$ is expressed as follows.

$$
x(t) = (v * h)(t) = \int v(t - \tau)h(\tau)d\tau
$$

It is necessary to consider the intra-subject correlation regarding the error term. Though several methods have been proposed, the pre-whitened method is frequently used. Matrix Σ, which becomes $\Sigma^{-1} = K^\top K$ by Cholesky factorization for the error correlation matrix. If you multiply this matrix from the left side of both sides of the expression (3.3),

$$
KY_i = KX\beta_i + K\varepsilon_i
$$

$\mathrm{Cov}(\boldsymbol{K\varepsilon}_i) = \boldsymbol{K}\mathrm{Cov}(\boldsymbol{\varepsilon}_i)\boldsymbol{K}^\top = \sigma_i^2\boldsymbol{I}$. Considering this as a new GLM, the theory in Section 3 of this chapter can be applied. Depending on the software, whether the 1st or 2nd level estimation is used also differs. For a more detailed account, please refer to Mumford and Nichols (2006) and Monti (2011). Behavior of the mixed effects model on fMRI data is examined by Miller et al. (2009) and points out that if the inter-group variation is greater than the intraspheric variation, it should be done with a smaller TR. In fMRI, measurement and analysis are performed more than once per subject per time, and a generalized linear mixture model is applied (Skup, 2010).

Multiple comparison and correction

In the example of GLM application provided in the previous section, a threshold value of the test statistic was set as "statistically significant", and only the statistic amount above the threshold was plotted. To make inferences in whole brain regions, multiplicity of verification considers multiple t-statistics (the number of voxels). When considering which of 100,000 voxels are significant, if the (unadjusted) significance level $\alpha = 0.05$, then 5,000 voxels will be false positives, and the likelihood of mistakes will be excessive. Thus, multiplicity correction is required (Nichols, 2012). Furthermore, as correlation between adjacent voxels can be strong, it must be considered. In this section, we first describe the correction method of multiplicity used in brain image analysis, and the following sections describe the consideration of correlation using cluster-level estimation and normal RFT (Gaussian RFT).

Why multiple correction

There is a major premise that errors occur in the test. When performing the test, the null hypothesis and the alternative hypothesis are set; the probability value (p-value) is calculated from the data and compared with a preset significance level (normally 5%, so the following is also assumed to be 5%). If the p-value is lower than the significance level, the null hypothesis is rejected (the alternative hypothesis is adopted) and the conclusion that the result is significant (such as a significant difference) is reached. If the p-value exceeds the significance level, the null hypothesis is not rejected. The test performed in this procedure is called a 5% significance level test. As a statistical precept, the p-value calculated from the data at hand is considered to have variations since the data to be analyzed is a finite sample (data of n people) obtained by random sampling. That is, if another sample were taken (usually one sample, which might be hard to imagine), the p-value from that sample may be different from that of the original sample. Assuming a situation where the p-value can take several values depending on the sample, not all will lead to a correct conclusion and a mistake may occur. The test controls the error so that it is small (it cannot be zero) before drawing conclusions.

Clinical trials emphasize the importance of error control. This is because statistical analysis is used as the last barrier for new treatments to emerge. This kind of analysis is called verification analysis and it strictly performs multiple corrections. There is another type of analysis called exploratory analysis that is often used in brain imaging analysis. The main purpose of exploratory data analysis is to find useful information from data analysis without setting a hypothesis to be verified. In many cases, the test result is not directly related to the conclusion and verification experiments are performed based on the results. Additionally, biological judgment is included and a comprehensive judgment is made. In such cases, it is accepted that some errors occurring in the test would be tolerated.

One test with a significance level of 5% can control up to 5% of the error probability. Similarly, we want to control errors up to 5% in the overall judgment obtained for multiple tests. However, if we perform multiple tests with a significance level of 5% multiple times, all those errors will be accumulated. Errors will occur with a higher probability than 5% (type 1 error rate inflation). The cause is that the original p-value and significance level calculated from each test were used and a "correction" that considers the overall misjudgment is required.

Before constructing a correction method, it is necessary to formalize errors in overall judgment, and a control method to reduce the mistake is only constructed subsequently. In practice, the correction is made, the larger p-value than the original calculated in each test by comparing it with the same significance level as in the case of performing one test to reduce the possibility of rejection by mistake. The p-value that has undergone correction is called the corrected p-value. Equivalently, the p-value is left as it is and the significance level is reduced by correction. This is called the adjusted significance level. A numerical example is given later. One of the widely used correction methods is the Bonferroni method. In the Bonferroni method the corrected p-values is obtained by multiplying each p-value by the number of tests. Additionally, the adjusted significance level is obtained by dividing the significance level by the number of tests. Thus, Bonferroni's method can be executed relatively easily from multiple p-values.

Such corrections tend to be difficult to reject. It is conservative to hesitate in rejecting many hypotheses (items) after rigorously correcting them. It may be effective when errors need to be tightly controlled, as in clinical trials. However, nothing may be obtained at the time of screening, for example, by making the correction strict in brain imaging analysis described above. In this case, being conservative is not necessarily desirable. For this reason, multiple corrections are not always considered.

Errors

There are two types of errors that can be made when deciding whether an effect is present. It is an error to declare that there is an effect when there is none; this is called a false positive. Similarly, when it is declared that there is no effect when there is one, it is called a false negative. In the test, false positives, which reject the null hypothesis although it is correct, are called type 1 error (alpha error), and false negatives, which do not reject the null hypothesis although the alternative hypothesis is correct, are called type 2 error (beta error). The probability that these errors will occur is called the error probability, and the value obtained by subtracting the beta error probability from 1 is called the power of the test. Additionally, type 1 and type 2 errors are competitive. That is, if the type 1 error is reduced, the type 2 error increases. The opposite is also true. Therefore, when constructing a test method, we try to keep one error constant and reduce the other error.

Theoretically, if the p-value is rejected by comparison with the significance level, the type 1 error probability is the same as the significance level. In other words, if a test with a significance level of 5% is performed (determining that the null hypothesis is rejected with a p-value of <0.05), the probability of a type 1 error can be reduced to 5%. The error probability remains at 5%, which is sufficiently low to decrease the likelihood of mistakes. It is only conventional that the 5% significance level is used.

In assessing effectiveness, the type 1 error is "finding an error" and its publication is scientifically misleading. The type 2 error would make it impossible to draw a scientific conclusion and it would be necessary to control the first kind of error. Thus, in most tests, the p-value threshold is set so that the type 1 error probability is controlled to 0.05 (the second-type error probability is neglected). Therefore, there is a possibility that the probability of error of the second kind is high; thus, even if the p-value is ≥ 0.05, it cannot be said that the null hypothesis is adopted.

Correction

An important point in performing multiple correction is determining the set of hypotheses (items to be tested) to be considered when calculating errors. A conclusion is made for each set. This set of hypotheses is called a family, for example, multi-group (combination), time measurement, evaluation item level, gene set, and voxel set. The error probability of a family unit is called a family-wise error (FWE). The setting of the family can be subjective and the result differs depending on the setting. Therefore, it is preferable to make a natural setting based on the background information of the data.

The following setting is considered to explain them. If some thresholds are set and V hypothesis tests are performed, a contingency table as shown in Table 3.1 can be expected. Each cell represents the number of corresponding tests. It is unknown whether the null hypothesis is true or false. Therefore, when the tests are performed, the information that can be known from Table 3.1 are the number of tests N_A for which the null hypothesis was adopted, the number of tests N_R for which the null hypothesis was rejected and the total V (the number of tests performed corresponding to the number of voxels in the brain imaging analysis).

Table 3.1: Contingency table for test results for whole voxel and true.

	Accept null hypothesis	Reject null hypothesis	sum
Null hypothesis is true	V_{0A}	V_{0R}	m_0
Null hypothesis is false	V_{1A}	V_{1R}	m_1
sum	N_A	N_R	V

Here are two representative correction methods.

(1) Family-wise Error Rate (FWER)

The FWE represents the presence of one or more false positives among multiple test results; in other words, at least one false positive is defined as causing a type 1 error and in Table 3.1, $V_{0R} > 1$. The family-wise error rate (FWER) is the probability. In cases where the number of tests is high and strong evidence is not required, when used as preprocessing in high-dimensional data analysis, a method that allows at least several false positives (for example, when V is 10 or more) is occasionally employed. This is called generalized FWER.

The FWE is the one or more false positives among multiple test results, while FWER is the probability that it occurs: i.e., FWER = Pr(FWE). Let T_k be the test statistic as given by the expression (3.2) ($k = 1, 2, \dots, V$). For the null hypothesis $H_0 : c'\beta_k = 0$, if $T_k \geq u$ for a constant u, H_0 is rejected. Then, FWER becomes as follows.

$$\text{FWER} = \Pr(\text{FWE}) = \Pr\left(\bigcup_k \{T_k \geq u\}|H_0\right) = \Pr\left(\max_k T_k \geq u|H_0\right)$$

Thus, FWER is expressed as a distribution of maximum values. In this way, FWER can be controlled as α by setting the maximum value distribution

$100(1 - \alpha)\%$ point as a threshold. That is,

$$\text{FWER} = \Pr\left(\max_k T_k \geq u_\alpha \mid H_0\right) = \alpha$$

where $u_\alpha = F_{max}^{-1}(1 - \alpha)$. F_{max} is the maximum value distribution and is obtained by the RFT in Section 3. See Nichols and Hayasaka (2003) for a more detailed account of the use of RFT-based FWER in MRI data analysis.

(2) False Discovery Rate (FDR)

As an alternative method, the concept of false discovery rate (FDR) has been developed in the field of high-dimensional data analysis. FDR is defined as FDR $= E[S/R]$ when R voxels become meaningful and S of them become an error—i.e., FDR is the ratio (expected value) of voxels rejecting H_0 when in fact it is true.

If a hypothesis test is conducted with some threshold value, a table like 3.1 can be created. The percentage of errors among rejected voxels is called False Discovery Proportion (FDP) and is expressed as follows.

$$\text{FDP} = \frac{V_{OR}}{V_{1R} + V_{OR}} = \frac{V_{OR}}{N_R}$$

However, if $N_R = 0$, FDP $= 0$. Since the truth of the null hypothesis is unknown (V_{OR} is unknown), consider the expected value to be FDR, that is, FDR $= E[\text{FDP}]$. Below, we explain the Benjamini and Hochberg (BH) method (Benjamini and Hochberg, 1995), which is a typical control method for the FDR.

When giving the threshold q (usually 0.05), the p-values are ordered by descending magnitude $p_{(1)} \leq p_{(2)} \leq \cdots \leq p_{(V)}$. Let i be the largest i that satisfies $p_{(i)} < i/V \times q$. The null hypothesis corresponding to $p_{(1)}, p_{(2)}, \ldots, p_{(r)}$ can then be rejected. The rejection in the BH method the value of FDR keeping it below q,

$$\text{FDR} = q\frac{m_0}{V} \leq q$$

The q-value (Storey, 2002, Storey, 2003) corresponding to the FDR-corrected p-value is called the q-value, which corresponds to $p_{(i)}$ is $q_{(i)} = \min_{m \geq i}(V p_{(i)}/m)$. For a more detailed account of the use of FDR in brain image data analysis, see Genovese et al. (2002).

In general, it is true that FDR $<$ FWER (FDR tends to detect more significant voxels than FWER, controlling the FWER implies that the FDR is also controlled), i.e., the control of FDR is less conservative than the FWER control (Benjamini and Hochberg, 1995). As in the case of high-dimensional data analysis, when the number of tests becomes substantial, controlling the FWER becomes conservative decision making; thus, FDR is practically more useful as a threshold decision. This will also be discussed in the next section. However, Goeman and Solari (2014) argue that it is substantially effortless for FWER regulation to be conservative and unsuitable for high dimensional data, such as in genetic research. A broader version of the FWER method is introduced in Dmitrienko and D'Agostino Sr (2013). As an alternative method of multiple correction, the permutation test (Nichols and Holmes, 2002) is executable with SnPM and explained later through the R code.

Commonly, to make multiple corrections is to change the p-value threshold (significance level) calculated by the members of the family. It is done so that the overall error probability (FWER, FDR) meets the prespecified level (e.g., 5%). Although there are several correction methods, it can be considered that only the threshold value (corrected significance level) is different. Conversely, if only the threshold value is desired to be kept, the p-values should be corrected (corrected p-value). The correction of the p-value threshold and the correction of the p-value itself are equivalent. Furthermore, the p-value is calculated from a value called the test statistic by using the probability distribution function. Thus, setting a threshold for the p-value is equivalent to setting a threshold for the test statistic.

Simple example for FWER and FDR correction

The calculation of FWER is explained as follows. Setting the significance level α, when the null hypothesis is correct, the probability that the null hypothesis is rejected is α; conversely, the probability that the null hypothesis is not rejected is $1 - \alpha$. When the test of the significance level α (= probability that the null hypothesis is rejected) is performed V times independently, the probability that all the null hypotheses are not rejected is $(1 - \alpha)^V$. Conversely, the probability of rejecting any one null hypothesis is

$$1 - (1 - \alpha)^V$$

This is (type 1) FWER. For example, if the test with significance level $\alpha=0.05$ is performed twice ($V = 2$),

$$1 - (1 - 0.05) \times (1 - 0.05) = 0.0975,$$

the FWER is 0.0975, which exceeds 0.05 (an error committed by the family). This is because the test with a significance level of 0.05 was repeated. If we increase the number of tests, the FWER increases, as shown in Figure 3.2. Thus, we should correct the FWER to 0.05. To correct it, do one of the following. (1) Set the significance level of the calculated p-values in the members of the family. (2) Conversely, if you want to keep the significance level, "correct" the

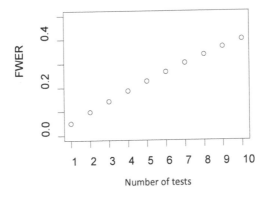

Figure 3.2: Type 1 FWERs for uncorrected independent tests.

p-values. There are many correction methods, but it can be considered that only the significance level (threshold) is different.

The simplest FWER correction method is Bonferroni correction and to apply it one of the following steps can be performed. (1) Divide the significance level by the number of tests (without p-values). (2) Multiply the number of tests by the p-value (with the significance level unchanged). In the previous example, each significance level 0.05 is halved (= 0.025), that is, the level is strict (it is difficult to reject the null hypothesis).

$$1 - (1 - 0.025) \times (1 - 0.025) = 1 - 0.975 \times 0.975 = 0.049375$$

This shows that Type 1 FWER could be controlled to 0.05. However, since this correction assumes independence, it tends to be conservative (= insufficient power), that is, fewer hypothesis are rejected and lower detentions.

The method for Bonferroni and FDR control will be explained using 10 artificially created p-values in Table 3.2 as numerical examples. The p-values in Table 3.2 are sorted in ascending order and the order is described. Consider a test that controls both FWER and FDR to 5%.

Bonferroni's method, as described earlier, is to multiply the resulting p-value by the number of tests or to divide the significance level by the number of tests. In Table 3.2, there are 10 hypotheses, so if the significance level of 0.05 is divided by each p-value and by the number of tests, the corrected significance level will be 0.005 and it is described in the "Threshold" column of the "Bonferroni" column in Table 3.2. Comparing this with each p-value, only the top two p-values (0.0025 and 0.0036) are below the adjusted significance level and are significant. If you multiply the p-value by the number of tests, the two smaller p-values are 0.025 and 0.036, which are the corrected p-values. This is compared to the original

Table 3.2: Corrected significance levels (thresholds) and tests by each method for the p-values of the 10 null hypotheses (H_0 = Null hypothesis).

Ordered p-value		Bonferroni (FWER 0.05)		BH (FDR 0.05)	
		Threshold	H_0	Threshold	H_0
1	0.0025	0.0050	Reject	0.0050	Reject
2	0.0036	0.0050	Reject	0.0100	Reject
3	0.0056	0.0050		0.0150	Reject
4	0.0075	0.0050		0.0200	Reject
5	0.0079	0.0050		0.0250	Reject
6	0.0124	0.0050		0.0300	Reject
7	0.0182	0.0050		0.0350	Reject
8	0.0420	0.0050		0.0400	
9	0.0520	0.0050		0.0450	
10	0.0600	0.0050		0.0500	

significance level of 0.05 and rejected. Obviously, one would obtain the same result as with the corrected significance level. This is a method of controlling the FWER.

Next, the Benjamin and Hochberg (BH) method will be described as a method of controlling the FDR. Given a controlled FDR (usually 0.05), sort the p-values in ascending order. The threshold in the BH method is 0.05×(rank of p-value/number of tests). Compare this with the p-value and reject the null hypothesis if the p-value is small. In the case of the example in Table 3.2, the 7 smallest p-values are rejected, and the FDR rejects more null hypotheses compared to FWER. The fact that the FDR has a higher detection power than the FWER also applies to the general case.

Finally, for clarity, the above contents are explained again in Figure 3.3. Here, the horizontal axis is the ascending order and the vertical axis is the p-value. The plotted • is the p-value corresponding to each null hypothesis. The dashed line is the significance level of 0.05 without correction, the dotted line is the corrected significance level of Bonferroni and the dashed and dotted lines are that of the BH method (a straight line with a slope of 0.05/10 = 0.005 and an intercept of 0). The Bonferroni method and the BH method reject the null hypothesis corresponding to the p-value below each line. Therefore, two null hypotheses are rejected by the Bonferroni method and seven by the BH method. If the area under each dotted line is large, it can be seen that the detection power is large. From this, it can be seen that the detection power of FDR is the highest among the correction methods and that the Bonferroni method is the most conservative.

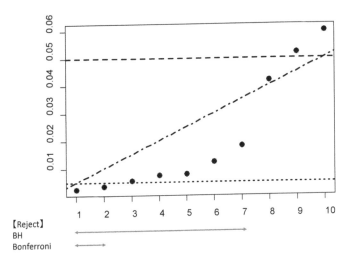

Figure 3.3: Bonferroni method and the BH method for Table 3.2.

Cluster inference

The number of tests increases as inference (voxel level inference) is performed for each voxel, and a higher threshold value is set (difficult to detect) even in multiple correction. Structurally, it is natural for neighboring voxels to have similar results and to consider a cluster of adjacent voxels. Speculation of targeting a cluster is called cluster level inference. The estimate for the number of clusters is called set level inference; however, as it is not often used, only cluster level inference is discussed herein.

Clusters are formed by the following procedure.

1. Compute the test statistics of each voxel and display them on the brain image template.

2. Decide an arbitrary threshold (height threshold) for the test statistic ($h = 2.5$ or 3). This is called cluster-defining threshold (CDT).

3. Form clusters from voxels exceeding the threshold clustering adjacent voxels as one cluster. SPM uses 18-connectivity.

Whether a cluster (voxel) is significant is judged based on the multiple corrected p-value. Consider the case like that of Figure 3.4. This is displayed on a color scale of six levels of statistics in a voxel of 8×8 (strictly speaking in the case of two dimensions). At the voxel level of Figure 3.4(A), a threshold for the value of the test statistic itself is obtained (only the darkest voxel is significant). At the cluster level of Figure 3.4(B), a certain threshold value is arbitrarily determined

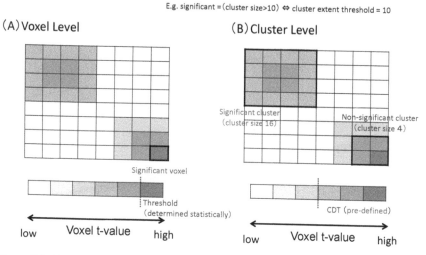

Figure 3.4: (A) Voxel level and (B) Cluster level. Clusters were assumed to be adjacent to two or more voxels that exceeded the threshold. For cluster-level inference, the threshold was set to 10 voxels.

for the test statistic (the third darkest or darker voxels) in order to determine the cluster. As a result, two clusters are formed. The threshold value for the cluster is considered to be the number of voxels belonging to the cluster. In the example of Figure 3.4, the threshold is set to 10 voxels, the cluster of 16 voxels is significant, and the cluster of 4 voxels is nonsignificant.

As for the cluster and voxel levels, multiplicity correction is required. The number of tests is relatively high in detection power because the cluster level is reduced. Chumbley et al. (2010) shows the effectiveness of the cluster level FDR (topological FDR) implemented in SPM 8. Although the threshold of topological FDR is the SPM's output, the setting of the corresponding extent threshold is required to display the result based on it. The statistical theory related to cluster inference is based on RFT in Section 3.

Random field theory

A random field (RF) is a set of ordered random variables. Random variables in brain image analysis are GLM test statistics obtained by the expression (3.2), and their ordering is considered to be the position of the voxel, and the theory is for determining the threshold after considering the correlation structure (Worsley, 2003). If the simultaneous distribution of the collection of random variables is a multivariate normal distribution, it is called Gaussian Random Field (GRF). The

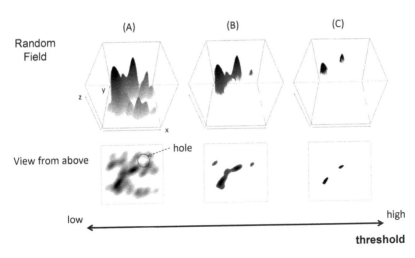

Figure 3.5: Relationship between threshold and random field. The threshold is moved in the z-axis (test statistic) direction of the same random field. In (A), there is one large cluster and one hole, In (B), there are three clusters, In (C), there are two clusters.

test statistic of each voxel is the t-statistic, but it can be regarded as following a normal distribution by z-transforming the p-value. In general, there are probability field theories for various distributions, but here we consider only the three dimensional GRF.

In Figure 3.5, for example, a certain random field is shown in the upper side, and a diagram from the top is shown on the lower side. Assume that spatial positions (voxels) are arranged in the directions of the x-axis (horizontal) and y-axis (longitudinal) and that the test statistic is shown in the z-axis (height) direction. As the threshold value in the z-axis direction is changed, clusters and the holes sandwiched between the clusters change from the top view of the random field.

R example

In this section, we explain the formation of clusters in a 1D Gaussian random field and how smoothing affects the cluster formation.

Original data and function

The example data is 1D Gaussian random filed. The number of coordinates was set to 20, numbered from 1 to 20, and a normal random number was generated at each coordinate and converted to an absolute value.

```
n = 20; x = 1:n
set.seed(1); y = abs(rnorm(n)); names(y) = x
```

This is the conversion function for the plot with different colors of above and below cdt.

```
cuty = function(y1, cdt){
rbind(ifelse(y1-cdt<0, y1, cdt),
ifelse(y1-cdt<0, 0, y1-cdt))
}
```

Differences with CDT

The clusters are plotted with the bar above the cdts (0.5 and 1.5) coloring.

```
par(mfrow=c(1, 2), mar=c(4,3,4,3))
for(cdt in c(0.5, 1.5)){
y2 = cuty(y, cdt)
barplot(y2, col=c(8,2), main=paste("CDT =", cdt))
abline(h = cdt, lty=2, col=2)
}
```

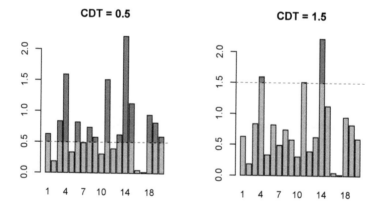

The smaller the CDT, the more clusters are likely to be created. On the other hand, when the number of clusters is considered to be too large, it is better to use a large CDT. However, as in this example, the values of adjacent coordinates are different and the probability fields with large variations require smoothing.

FWHM and cluster above CDT

For smoothed data, the clusters are plotted with different FWHM values (2 and 8). The CDT is fixed at 0.75.

```
cdt = 0.75

par(mfrow=c(1,2), mar=c(4,3,4,3))
for(f1 in c(2,8)){
sy = gsmooth(x, y, f1); y2 = cuty(sy, cdt)
barplot(y2, col=c(8,2), main=paste("FWHM =", f1), ylim=c(0,2))
abline(h = cdt, lty=2, col=2)
}
```

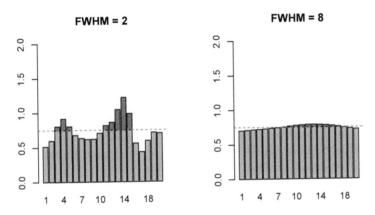

As shown in the figure, the low smoothness (FWHM = 2) yields small cluster sizes, a large number of clusters, and narrow clusters. On the other hand, a high (FWHM = 8) yields large cluster sizes, a small number of clusters, and wider cluster width.

The Euler characteristic χ_h plays an important role in the RFT and can be expressed as χ_h = (number of clusters) - (number of holes). As shown in Figure 3.5, the Euler characteristic depends on the CDT h because the number of clusters and the number of holes are determined by setting the threshold value. FWER is expressed as an expected value of the Euler characteristic as follows.

$$\text{FWER} = \Pr(\max_k T_k \geq h|H_0) = \Pr(\text{more than one cluster }|H_0)$$
$$\approx \Pr(\chi_h \geq 1|H_0) \approx E[\chi_h|H_0]$$

The first approximation of the second line is based on the null hypothesis that "there is no hole" and only considering the cluster. In the case of three dimensional GRF, the expected value of the Euler characteristic is given as follows (Worsley et al., 1996).

$$E[\chi_h] = R\frac{(4\log 2)^{3/2}}{(2\pi)^2}(h^2 - 1)\exp\left(-\frac{h^2}{2}\right) \tag{3.5}$$

However, R is called the resolution element (RESEL) per unit of smoothness and is given as follows.

$$R = \frac{\text{Vol}(\Omega)}{r} \text{ where } r = \text{FWHM}_x \times \text{FWHM}_y \times \text{FWHM}_z \tag{3.6}$$

$\text{Vol}(\Omega)$ is the volume of Ω, $\Omega \in \mathbb{R}^3$ is the area to be analyzed, r is the RESEL, and FWHM_ℓ is FWHM of the peripheral density function in the ℓ direction of the random field. $\ell = x$ (horizontal), y (vertical), z (slice). "Smooth" means that the random variable of the random field is small. R represents the smoothness of the probability field, and $E[\chi_h]$ depends on the smoothness of the random field. For the expression (3.6), $\text{Vol}(\Omega)$ uses the integral of the density function. The FWHM estimation method is described below.

For GRF, the FWHM estimator is given by $\text{FWHM}_\ell = \sqrt{8\log 2W_{\ell\ell}}$ ($\ell = x, y, z$), where $W_{\ell\ell}$ is a diagonal element of the matrix $W = (2\Lambda)^{-1}$, Λ is the variance covariance matrix of partial derivatives with respect to the three dimensional direction of GRF in voxels and represents the roughness (inverse of smoothness) of GRF. Since the matrix Λ is unknown, it is estimated from the residuals of the GLM in the expression (3.1) (Kiebel et al., 1999).

The (standardized) residual of voxel k is as follows.

$$\tilde{e}_k = \frac{e_j}{\sum_j e_j^\top e_k}, \quad e_k = Y_k - X\hat{\beta}_k$$

The diagonal elements of the matrix Λ are estimated by the following equation (off-diagonal elements are assumed to be 0).

$$\hat{\lambda}_{\ell\ell} = \lambda_v \frac{v-2}{(v-1)N}\sum_{k\in\tilde{\Omega}}\sum_{j=1}^{n}\frac{\partial\tilde{e}_{jk}}{\partial\ell}$$

where n is the number of subjects, $\tilde{\Omega}$ is the set of voxel addresses representing the area to be analyzed, N is the total number of voxels, and v is the effective degree of freedoms: it is given by $v = \{\text{trace}(S)\}^2/\text{trace}(S^2)$ where $S = X(X^\top X)^{-1}X^\top$. Partial differentiation of the residual is due to the difference between adjacent voxels. λ_v is a normalization constant and is given as follows.

$$\lambda_v = \int_{-\infty}^{\infty}\frac{(t^2+n-1)^2}{(v-1)(v-2)}\frac{\psi_v(t)^3}{p(t)^2}dt$$

where $p(t) = \phi(\Phi^{-1}(1 - \Psi_v(t)))$, ϕ and Φ are standard normal density and distribution functions, respectively. ψ_v and Ψ_v are v degrees of freedom t-distribution density function and distribution function, respectively. In this way, the expected value of the Euler characteristic of equation (3.5) is estimated from the data, and the FWE multiple correction of the p-value based on it is performed at the voxel level and the cluster level as follows.

The FWE corrected p-value according to voxel level RFT for the statistic t calculated from the data in a certain voxel is given as follows.

$$p_{\text{Voxel FWE corr}} = R\frac{(4\log 2)^{3/2}}{(2\pi)^2}(t^2 - 1)\exp\left(-\frac{t^2}{2}\right)$$

From this equation, if the random field is smooth, R becomes smaller and the correction becomes lighter. As the target volume becomes larger, R becomes larger and the correction becomes severe.

Consider cluster level inference. Since it is based on the cluster size (the number of voxels in the cluster), first consider cluster size. The expected value of the cluster size S is given as $E[S] = E[N]/E[L]$. However, N is the total volume of the area exceeding the threshold value and is given by $E[N] = \text{Vol}(\Omega)\text{Pr}(T > h)$. T is a random variable that follows the same distribution as the t-test statistic. L is the number of clusters and is represented by the Euler characteristic—i.e., $E[L] \approx E[\chi_h]$. Thus, the FWE-corrected p-values according to cluster-level RFT are given as follows (Friston et al., 1994).

$$p_{\text{Cluster FWE corr}} = 1 - \exp(-E[\chi_h]p_s)$$

where p_s is an uncorrected cluster-level p-value and is as follows.

$$p_s = Pr(S > s) = \exp(-\theta s^{2/3}) \text{ where } \theta = \left\{\frac{\Gamma(5/2)E[\chi_h]}{\text{Vol}(\Omega)Pr(T > h)}\right\}^{2/3} \quad (3.7)$$

Thus, in order to make an inference based on RFT (to obtain the p-value), (1) the total volume of the part to be analyzed and (2) smoothness (FWHM if tracing backward) are necessary. Smoothness of the test statistic is obtained if the original voxel value is smooth. In order to lighten multiple correction, image smoothing is performed as preprocessing. For the cluster estimation based on the random field, see references Worsley et al. (2004) and Ashby (2019).

Simple example for cluster size test

Figure 3.6 shows the GLM results from the setting in Figure 3.1. Figure 3.6(A) shows the T-map in which the test statistic (T value) calculated for each voxel is shown on the brain image template as the result of the t-test performed for each

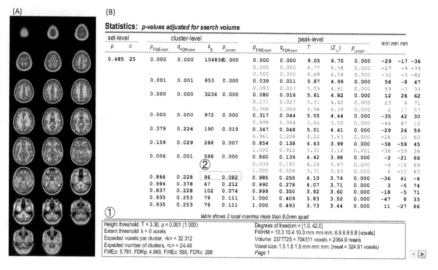

Figure 3.6: Result for SPM (A) T-map (B) Statistics table.

voxel (threshold value is not set). If you set some value, you can obtain a table of statistical analysis results as shown in Figure 3.6(B).

Multiple large and small clusters can be obtained from the entire brain image and whether or not a cluster (voxel) is significantly large is judged based on the p-value. The cluster size is the number of voxels that form each cluster. The value of k_E at the cluster level in Figure 3.6(B) represents several cluster sizes obtained with CDT = 3.3. The null hypothesis in the test is "cluster size = 0." If the null hypothesis is rejected because the calculated p-value is smaller than the significance level, the cluster is judged as being significantly large. Random field theory is used to derive the p-value calculation formula. The assumption is that the brain image test statistic is smooth. Therefore, smoothing must be performed as a preprocessing step.

The equation (3.7) can be approximated and simplified as follows. From equations (3.5) and (3.6), using the normal density function $\phi(h) = \exp(-h^2/2)(2\pi)^{-1/2}$

$$E[\chi_h] = \frac{\text{Vol}(\Omega)}{r}(4\log 2/2\pi)^{3/2}(h^2 - 1)\phi(h) = \frac{\text{Vol}(\Omega)}{r}a(h^2 - 1)\phi(h) \quad (3.8)$$

where $a = (4\log 2/2\pi)^{3/2}$. Assuming the standard normal distribution of the approximation of the t-distribution as the distribution of T and using the Mills ratio, $Pr(T > h)$ can be written as $Pr(T > h) = 1 - \Phi(h) \approx \phi/h$. From these and the equation (3.8), and with $b = (a\Gamma(5/2))^{2/3}$,

$$\theta = \left\{ \frac{\Gamma(5/2)\frac{\text{Vol}(\Omega)}{r}a(h^2-1)\phi(h)}{\text{Vol}(\Omega)\phi/h} \right\}^{2/3} = b\left\{ \frac{1}{r}h(h^2-1) \right\}^{2/3}$$

Thus, the approximated cluster-level p-value is given by

$$p_s \approx \exp\left(-b\left\{ \frac{s}{r}h(h^2-1) \right\}^{2/3} \right) \tag{3.9}$$

where $b = (4\log 2/2\pi)\Gamma(5/2)^{2/3}$ and a constant $b=0.5334942$.

h represents CDT and r represents RESEL. RESEL is calculated as the product of FWHM in three directions. Note that the FWHM here is for the T-map and is determined by the residual of GLM, unlike the one set in the preprocessing. As shown in the right column of Part (1) in Figure 3.6(B), the FWHM in the three directions is calculated as 6.9, 6.9, and 6.9. The value that is displayed as resel = 324.91 voxels as the product of the two lines below and it is rounded off and subsequently used as $r = 325$. Equation (3.9) is relatively simple and can be calculated with a simple computer. If $p_s < 0.05$, the significance level is 0.05 (5%), and clusters with cluster size can be judged to be significantly large.

Furthermore, solving equation (3.9) for s yields the following equation:

$$s = \frac{r[-\log p_s/b]^{3/2}}{h(h^2-1)} \tag{3.10}$$

Substituting a significance level (for example, 0.05) into p_s in this formula, the corresponding cluster size is calculated, which is the cluster size threshold (size threshold) for the assigned significance level, and clusters larger than that threshold are said to be at the significance level and can be considered significant.

An example of using this formula for the T-map in Figure 3.6(A) is shown. The test of $s = 96$ is performed with $b = 0.53$, $h = 3.3$, $r = 325$. Substituting a value into equation (3.9)

$$p_s = \exp(-0.53 \times [96/325 \times 3.3 \times (3.3^2 - 1)]^{2/3}) = 0.093$$

It can be seen that a value close to 0.082 of the SPM output is calculated. For reproducibility, b is rounded up and calculated. However, if b is calculated using the correct value, $p_s = 0.089$, which is closer to the SPM output. In this situation, the size threshold at the significance level of 0.05 is obtained. Substituting a value into Equation (3.10)

$$s = \frac{325 \times [-\log 0.05/0.53]^{3/2}}{3.3 \times (3.3^2 - 1)} = 133.8$$

This means that cluster sizes greater than 134 are significant at the significance level of 0.05. This can be entered as 134 in the SPM extent threshold. Figure 3.7 shows a significant region for the cluster size threshold s (it is denoted as k in the figure) calculated from the formula (3.10) for h=3.3, h=4.52 and 5.02 (equivalent to uncorrected p = 0.001, 0.000025 and 0.000001, respectively) in the T-map in Figure 3.6.

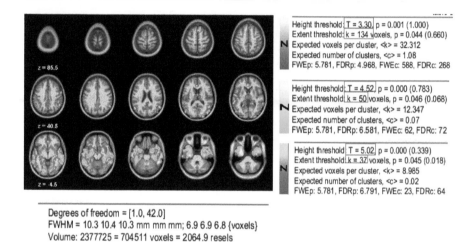

Height threshold T = 3.30, p = 0.001 (1.000)
Extent threshold: k = 134 voxels, p = 0.044 (0.660)
Expected voxels per cluster, <k> = 32.312
Expected number of clusters, <c> = 1.08
FWEp: 5.781, FDRp: 4.968, FWEc: 588, FDRc: 268

Height threshold T = 4.52, p = 0.000 (0.783)
Extent threshold k = 50 voxels, p = 0.046 (0.068)
Expected voxels per cluster, <k> = 12.347
Expected number of clusters, <c> = 0.07
FWEp: 5.781, FDRp: 6.581, FWEc: 62, FDRc: 72

Height threshold T = 5.02, p = 0.000 (0.339)
Extent threshold k = 37 voxels, p = 0.045 (0.018)
Expected voxels per cluster, <k> = 8.985
Expected number of clusters, <c> = 0.02
FWEp: 5.781, FDRp: 6.791, FWEc: 23, FDRc: 64

Degrees of freedom = [1.0, 42.0]
FWHM = 10.3 10.4 10.3 mm mm mm; 6.9 6.9 6.8 {voxels}
Volume: 2377725 = 704511 voxels = 2064.9 resels
Voxel size: 1.5 1.5 1.5 mm mm mm; (resel = 324.91 voxels)

Figure 3.7: Significant cluster at 5% level (CDT = 3.3, 4.5, 5).

In Figure 3.7, the value of p in the height threshold line indicates the significance level, and a value close to 0.05 is calculated, indicating that equation (3.10) is well approximated. A function called Transform SPM-maps of CAT12 is useful for setting the threshold.

R example

Cluster Level Inference

```
exp(-0.53*(96/325*3.3*(3.3^2-1)^(2/3)))
```

```
## [1] 0.09251765
```

```
s=96; h=3.3; r=325;
Es=(4*log(2))^(3/2)*h*(h^2-1)/(2*pi)^(3/2);
d=(gamma(5/2)*Es)^(2/3);
exp(-d*(s/r)^(2/3))
```

```
## [1] 0.08922926
```

```
(325*(-log(0.05)/0.53)^(3/2))/(3.3*(3.3^2-1))
```

```
## [1] 133.8178
```

Cluster p-value

We illustrate how the p-value changes as FWHM changes for cluster sizes s = 1, 10, and 50. The CDT is fixed at 4.

```
ss = c(1, 10, 50)
h = 4
b = (4*log(2)/(2*pi))*(gamma(5/2))^(2/3)
fs = 2:10
```

```
par(mar=c(2,2,2,6))
for(sidx in 1:length(fs)){
p = exp(-b*(ss[sidx]/fs^3*h*((h)^2-1))^(2/3))
if(sidx == 1){
plot(fs, p, lty = sidx, type="l", ylim=c(0,1),
xlab="FWHM", ylab="p-value", main="")
}else{
points(fs, p, lty = sidx, type="l")
}}
abline(h=0.05, lty=4, lwd=2)
par(xpd=TRUE)
legend(par()$usr[2], par()$usr[4],
legend=paste("s =", ss[order(ss)]), lty=order(ss))
```

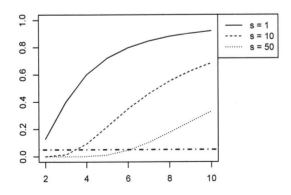

In this way, it appears that the power is higher if it is not smooth, but it should be noted that if it is not smooth enough, we can only obtain clusters with small cluster sizes. Here we have an extreme s=1 example, which is not significant by 5% in any FWHM. Since too large a FWHM is unlikely to be significant, an appropriate value for the FWHM should be set in the pretreatment. This will be discussed later, but 8mm is recommended, which is also the default for SPM.

TFCE

The cluster size test method traditionally used in SPM is a test method in which a CDT is applied to a T-map from a smoothed image and the result is highly selection sensitive. In contrast, a method that does not depend on the CDT has been proposed. Smith and Nichols (2009) proposed a threshold-free cluster enhancement (TFCE) method. Comparative studies in Li et al. (2017) showed the usefulness of the TFCE method. Raffelt et al. (2015) has been extended to the TBSS framework as a connectivity-based fixel enhancement (CFE). Lett et al. (2017) extends to cortical analysis. The TFCE method can be implemented in FSL and CAT12.

The TFCE method is explained using artificial data assuming a T-map from a two-dimensional random field as shown in Figure 3.8. There are four peaks, from the left, a high peak in a narrow area, a low peak in a wide area, and two adjacent peaks.

T-map

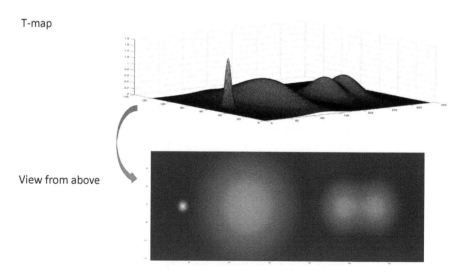

View from above

Figure 3.8: Two-dimensional random field.

TFCE starts by first transforming the original random field. The TFCE output at voxel v is given by

$$TFCE(v) = \int_{h_0}^{h_1} e(h,v)^E h^H dh \qquad (3.11)$$

where h_1 is the highest threshold and h_0 is the lowest threshold, typically $h_0 = 0$, and E and H take values of 0.5 and 2, respectively. $e(h,v)$ is the size of the cluster containing voxel v at threshold h.

This integral calculation is approximated in the form of a sum after preparing the threshold h discretely as a sequence. Calculate TFCE in the previous example, assuming that $E = 0.5$ and $H = 2$, for the sake of simplicity, h is considered in only four states: $h = 0.2, 0.4, 0.6$ and 0.8 (Figure 3.9). In practice, you will use many sequences with smaller steps.

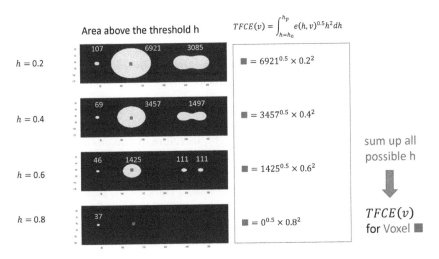

Figure 3.9: Process of TFCE calculation.

First, when $h = 0.2$, the two adjacent peaks on the right became one cluster due to the low threshold and the cluster became three in total. Each size (the number of pixels in the cluster) is 107, 6921, 3085. This calculates $e(h,v)^E h^H$ for the pixel v represented by the square. The cluster to which this pixel belongs has a size of 6921 and

$$e(h,v)^E h^H = 6921^{0.5} \times 0.2^2$$

Next, when $h = 0.4$, the area above the threshold is smaller than that above the threshold of $h = 0.2$. The Size of the cluster to which the pixel represented by the square belongs is 3457. Accordingly, the calculation of $e(h,v)^E h^H$ changes as shown in Figure 3.9.

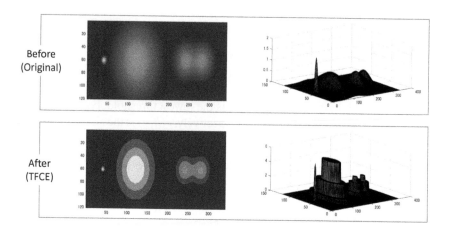

Figure 3.10: TFCE map.

By doing the same for $h = 0.6$ and $h = 0.8$, we get $e(h,v)^E h^H$ at each h and add them together to get TFCE(v) at v. The lower part of the Figure 3.10 shows the result when this is performed for all pixels.

Since the leftmost peak is a small area, simple cluster estimation is not likely to be significant. On the other hand, since the TFCE reflects both the size and height of the cluster, this peak has a height; therefore, it remains as a high value in TFCE. Next, the second peak from the left is the largest TFCE by reversing the left peak in TFCE because of the large cluster. The remaining two peaks are of medium height and large, so they remain intact. Thus, TFCE generally converts the original voxel value (test statistic) into a voxel value that reflects the size and height of the cluster to which it belongs and after converting it to TFCE, converts it to a new test statistic (per voxel) to be evaluated statistically. In many cases, the permutation test (multiple correction) is used.

R example

The following function computes the integrated values of the threshold-free cluster enhancement (TFCE) at each cdt.

```
TFCE0 = function(y1, cdt, E=0.5, H=2){
cy = 1*(y1 >= cdt)

clsta0 = c(1, as.numeric(names(which(diff(cy)!=0))))
clend0 = c(clsta0[-1]-1, length(cy))
clsta = clsta0[cy[clsta0]==1]
clend = clend0[cy[clsta0]==1]
```

```
clustidxs = lapply(1:length(clsta),
function(i) clsta[i]:clend[i])
clustsize = unlist(lapply(clustidxs, length))

x = 1:length(y1)
clust = unlist(lapply(x, function(x1){
a = which(unlist(lapply(clustidxs,
function(cidx) x1 %in% cidx)))
ifelse(length(a)>0, a, NA)
}))

a = clustsize[clust]^E * cdt^H
ifelse(is.na(a),0, a)
}
```

This function supports the plot.

```
TFCE1 = function(f1){
for(cdt1 in cdts2){
sy = gsmooth(x, y, f1)
y2 = cuty(sy, cdt1)
barplot(y2, col=c(8,2), main=paste("CDT =", cdt1))
abline(h = cdt1, lty=2, col=2)
tfce = TFCE0(sy,cdt1)
barplot(tfce, main=paste("TFCE CDT =", cdt1), ylim=c(0,10))
}
}
```

TFCE process

The TFCE procedure process is explained with CDTs $= 0.75, 0.5$ and 0.25, and FWHM $= 2$ in the following code.

```
cdts2 = c(0.75, 0.5, 0.25)
par(mfrow=c(length(cdts2), 2), mar=c(3,4,2,4))
TFCE1(f1=2)
```

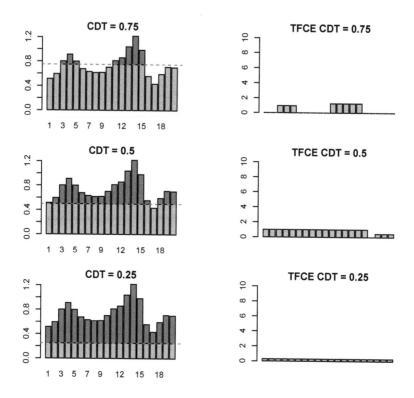

The values in the right column are computed for each coordinate, but if they belong to the same cluster, the values (the heights of the bars) are equal.

FWHM and TFCE

The different FWHM values produce the different TFCEs and the moderate FWHM value is preferable.

```
f1s = c(0, 2, 4, 8)
par(mfrow=c(length(f1s),2), mar=c(3,4,2,4))
for(f1 in f1s){
if(f1 == 0){barplot(y, main="Original")}else{
sy = gsmooth(x, y, f1)
barplot(sy, main=paste("FWHM =", f1), ylim=c(0,2))
}
if(f1 > 0){ sy = gsmooth(x, y, f1) }else{ sy = y }
cdts = c(0, sort(unique(sy)))
tfce = colSums(do.call(rbind,lapply(cdts, function(cdt1){
TFCE0(sy,cdt1)
```

```
})))
barplot(tfce, main=paste("TFCE FWHM =", f1))
}
```

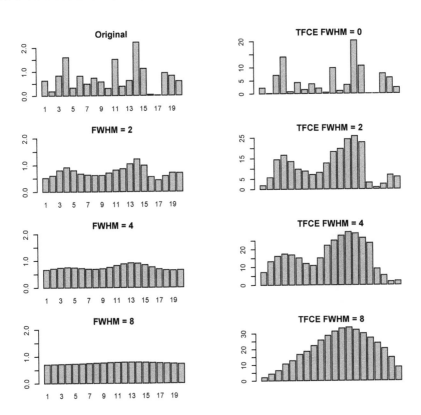

TFCE is also affected by smoothing range; when the FWHM is large, a large area is likely to be detected, and conversely, when the FWHM is small, there is more variability, making it difficult to obtain useful findings. Therefore, it is necessary to set an appropriate value of FWHM in preprocessing in the case of TFCE as well.

Permutation test

In normal hypothesis testing, the test statistic is constructed on the assumption that the null hypothesis is correct and the distribution that the statistic follows is determined theoretically. This is called a null distribution. Then, a threshold or p-value for determining whether to reject the null hypothesis or not is calculated from the test statistic using the null distribution.

An appropriate distribution can be considered to be a null distribution. The binomial distribution is often used in the analysis of binary data, but the subjects are independent of each other and the variance is determined by the sample size and the probability of occurrence. The binomial distribution is also approximated by a normal distribution when the sample size is large. The normal distribution is often used for continuous data. If the sample size of data is sufficient, the sample mean follows a normal distribution; otherwise, the data itself must follow a normal distribution. If the variance is known, it follows the normal distribution, but if it is unknown, it follows the t distribution. Using theoretical distributions like these requires assumptions about the data. It is difficult to verify rigorously the assumption and accuracy of data needed to use the method with confidence. Additionally, the difficulty is expected to increase for data with many variables as in multivariate analysis and in multiple comparison.

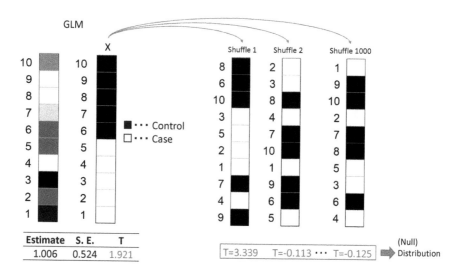

Figure 3.11: Example for permutation in GLM.

Thus, a test method that does not depend on the theoretical distribution is useful and the permutation test performs it. The permutation test creates a null distribution using random numbers generated by a computer and performs a statistical hypothesis test based on it. Here, GLM is used for explanation in Figure 3.11. Suppose there is one voxel value for a total of 10 cases, 5 cases and 5 control cases. In the GLM, Y is a voxel value and the design matrix X is a binary value representing a group of Case or Control, however, the intercept is omitted. In this example, the regression coefficient (mean value difference) is 1.006, the standard error is 0.524 and the test statistic T = 1.921 is calculated. Here, the voxel value

Y is left as it is and the order of the subjects in X is randomly changed. The resulting design matrix is Shuffle 1. Calculating the test statistic again from this and the GLM for Y yields T = 3.339. By repeating the random replacement 1000 times, 1000 corresponding test statistics can be obtained.

If the null hypothesis is correct, the test statistic is likely to be 0, because there is no difference between the mean values of Y in the cases and controls, but the values are slightly shifted due to random samplings (of finite size). This results in variations. The random sampling is simulated by shuffling. If the null hypothesis is correct, the value of the test statistic is expected to be 0 even if the replacement is performed and the error will vary. The distribution obtained by repeating the permutation is shown as a histogram and the rejected area is expressed as a percentage point (rank from the top). The details are explained using the R code in the next section.

R example

First, the data is prepared.

```
n = 5; x0 = rep(c(0,1), each = n)
set.seed(1); y = ifelse(x0==0, rnorm(n, 0), rnorm(n, 1))
```

Assuming that the number of data in one group is 5, the dummy variables with x0 = 0 for the control and x0 = 1 for the case are used as explanatory variables. For the objective variable, a random number is generated from a normal distribution such that the control average is 0 and the case average is 1.

Next, prepare a function to calculate the test statistic of the t-test.

```
fitfunc = function(idx){
x = x0[idx]
fit = summary(lm(y~x))
coef(fit)[2,1:3]
}
```

The argument idx specifies the order of dummy variables that are explanatory variables. By replacing idx with 1,2,3,...,10, the analysis result for the original data without replacement can be obtained. Regression analysis is performed using the explanatory variable and the objective variable specified by idx. That is, the design matrix consists of two columns: an intercept and a group dummy variable. Then, the estimated values of the regression coefficient, standard error, and test statistic are output.

Calculate the test statistic of the t-test from the data. idx is set to 1,2,3,...,10.

```
(fit0 = fitfunc(1:(2*n)))
```

```
##   Estimate Std. Error   t value
## 1.0058658  0.5236289  1.9209517
```

This is the estimation result for the original data. To make a statistical guess for this result, we replace the dummy variables and find the null distribution.

Prepare a randomly replaced index and calculate the test statistic of the t-test again. By resetting the seed of the random number, the test statistic for three types of permutations are calculated.

```
set.seed(1); idx = sample(2*n); round(fitfunc(idx), 3)
```

```
##   Estimate Std. Error   t value
##      1.366      0.409     3.339
```

```
set.seed(2); idx = sample(2*n); round(fitfunc(idx), 3)
```

```
##   Estimate Std. Error   t value
##     -0.071      0.632    -0.113
```

```
set.seed(1000); idx = sample(2*n); round(fitfunc(idx), 3)
```

```
##   Estimate Std. Error   t value
##     -0.079      0.632    -0.125
```

The above steps are repeated to output only test statistics.

```
tststs = sapply(1:1000, function(s1){
set.seed(s1); idx = sample(2*n); fitfunc(idx)[3]
})
```

1000 test statistics are obtained. Calculate the 97.5% quantiles (the two-sided significance level of 5% rejection).

```
(q1 = quantile(tststs, prob=0.975))
```

```
##     97.5%
## 2.231768
```

```
ifelse(abs(fit0[3])>q1, "reject", "not reject")
```

```
##       t value
## "not reject"
```

The test statistic before the swap was 1.921 (< 2.232); therefore, the null hypothesis was not rejected. From the t-distribution density function with eight degrees of freedom, which is the theoretical distribution, the rejection region is obtained by integrating the density function.

```
(q2 = qt(0.975, 2*n-2) )
```

```
## [1] 2.306004
ifelse(abs(fit0[3])>q2, "reject", "not reject")
```

```
##        t value
## "not reject"
```

In this example, the "t" value is 2.306, which is almost the same as that obtained by the replacement.

In the figure, the test statistic obtained by 1000 permutations is represented as a histogram. The figure also plots the t distribution density function with eight degrees of freedom, which is the theoretical distribution.

```
hist(tststs, xlab="T stat", main="", freq=FALSE,
ylim=c(0, 0.4))
abline(v=fit0[3], col=2, lty=2)
abline(v=q1, col=4, lty=3); abline(v=q2, col=3, lty=4)
f1 = function(x)dt(x, 2*n-2)
curve(f1, -5, 5, add=TRUE, col=3)
```

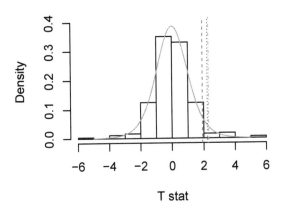

The test statistic obtained from the data is indicated by a dashed line, quantiles from the permutation distribution are indicated by a dotted line, and quantiles from the theoretical distribution are indicated by a dashed and dotted line. It also seems to fit the histogram obtained by permuting. It can be seen that the rejection area is wider in the permutation test (the quantile is smaller).

Next, the p-value is calculated. The p-value is the percentage of values obtained by swapping values that are higher than the test statistic 1.921 and is doubled on both sides.

```
(pp = 2*(mean(tststs > fit0[3])))
```

```
## [1] 0.116
```

In this case, it is 0.116 and the null hypothesis is not rejected at the significance level of 5%.

We compute the p-value of the t-test for comparison.

```
(tp = 2*(1 - pt(fit0[3], 2*n-2)))
```

```
##    t value
## 0.0909819
```

From this theoretical distribution, the p-value is 0.0909819. In this example, the theoretical distribution is a more conservative test.

To obtain a more reliable distribution, a larger number of samples is needed. The number of samples corresponds to the number of permutations. Here, a simple experiment was performed as follows.

```
nreps = round(c(seq(10, 100, length=10),
seq(200, length(tststs)-100, length=10)))
q1s = sapply(nreps, function(x) {
sapply(1:1000, function(y) {
set.seed(y); quantile(tststs[sample(1:length(tststs), x)],
0.975)})})
colnames(q1s)=nreps
```

From the test statistic obtained by the 1000 permutations, the number of permutations is randomly taken to calculate the quantiles. There were one thousand extractions.

A box plot was created with the number of permutations on the horizontal axis and the quantiles on the vertical axis.

```
boxplot(q1s, main="", xlab="Number of permutations",
ylab="Two-sided 95% point", type="b")
abline(h=q2, col=3, lty=2)
```

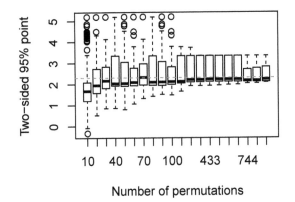

Number of permutations

As the number of permutations increases, the variation in quantiles is small and stable. In this way, a reliable distribution can be obtained by increasing the number of replacements. However, as the number of replacements increases, more calculation time is required and the more it needs to be accounted for.

Permutation based multiple correction

If the statistic is simply the difference between the mean values, as in the example, then the theoretical distribution can easily be obtained as if it were a t-distribution. If the statistics are complicated, such as in the cases of increasing the number of columns in the design matrix and in multiple comparison correction (with correlation), the theoretical distribution includes an approximate derivation, and its worst fitness to the data may be occur. In such cases, the permutation test may be effective. We will explain multiple comparison correction in cluster inference using R code.

R example

A cluster is formed from a one-dimensional random field that exceeds a certain threshold (= CDT) and the size of the cluster is tested. Multiple corrections are necessary because there can be multiple clusters. One way to control FWER is to consider the distribution of the maximum values of multiple statistics. Here, the method is explained. First, the function to be used should be prepared.

```
clustsize = function(y1, cdt){
cy = 1*(y1 >= cdt)

if(all(cy==0)){
```

```
clustsize = 0
}else{
clsta0 = c(1, as.numeric(names(which(diff(cy)!=0))))
clend0 = c(clsta0[-1]-1, length(cy))
clsta = clsta0[cy[clsta0]==1]
clend = clend0[cy[clsta0]==1]

clustidxs = lapply(1:length(clsta),
function(i) clsta[i]:clend[i])
clustsize = unlist(lapply(clustidxs, length))
}
clustsize
}

gsmooth = function(x, y, FWHM){
sigma = FWHM / 2*sqrt(2*log(2))
sy = sapply(x, function(x1)
weighted.mean(y,
dnorm(x, x1, sigma)/sum(dnorm(x, x1,sigma))) )
names(sy) = x
sy
}

cuty = function(y1, cdt){
rbind(ifelse(y1-cdt<0, y1, cdt),
ifelse(y1-cdt<0, 0, y1-cdt))
}
```

The next step is to set up the data. The number of coordinates in the random field is 20. The random field generated by normal random numbers is smoothed with FWHM = 2. From there, CDT = 0.75 forms adjacent coordinates as clusters.

```
n = 20; x = 1:n; f1 = 2; cdt1 = 0.75
```

Next, calculate the cluster size (the number of coordinates) as well as the maximum value. Repeat this process four times.

```
sidxs = 1:4
par(mfrow=c(length(sidxs),1), mar=c(3,4,2,4))
for(sidx in sidxs){
sigma1 = ifelse(sidx==1, 1, 0.9)
set.seed(sidx); y = abs(rnorm(n,0,sigma1)); names(y) = x
sy = gsmooth(x, y, f1); y2 = cuty(sy, cdt1)
csize = clustsize(sy, cdt1)
```

```
barplot(y2, col=c(8,2),
main=paste("Max Cluster Size =", max(csize)), ylim=c(0,1.2))
abline(h = cdt1, lty=2, col=2)
}
```

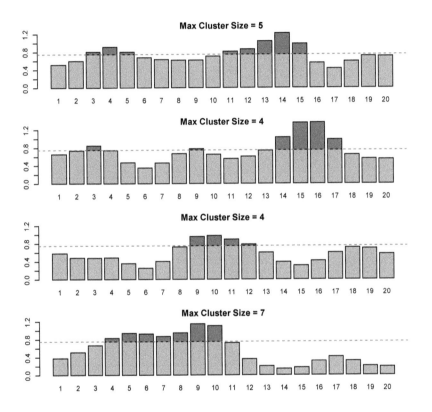

The result is the resulting figure. In the first row, there are two clusters, the sizes of which are 3 and 5, so the maximum value is 5. By doing the same in the second to the fourth row, the maximum values of the cluster sizes are calculated as 4, 4 and 7 as shown at the top of the figure.

Here, random numbers were generated as 20 test statistics and by repeating the random numbers, four different random fields were generated. Assume a random field with 20 test statistics obtained by transposing the GLM design matrix as seen earlier. That is, as the maximum cluster size is observed in the first row, the null distribution is created by the second and subsequent cluster size maximums. This time, the following code is used to create a distribution with 1000 repetitions.

```
sidxs2 = 1:1001
maxcsizes = sapply(sidxs2, function(sidx){
sigma1 = ifelse(sidx==1, 1, 0.9)
set.seed(sidx); y = abs(rnorm(n,0,sigma1)); names(y) = x
sy = gsmooth(x, y, f1)
csize = clustsize(sy, cdt1)
max(csize)
})
(obs = maxcsizes[1])
```

```
## [1] 5
```

```
(q1 = quantile(maxcsizes[-1], 0.95))
```

```
## 95%
##  11
```

The resulting histogram is shown here. The 95% quantile is 11, which is the rejection threshold. In this case, the result is that the null hypothesis is not rejected because the maximum cluster size calculated from the data was 5 ($<$ 11).

```
hist(maxcsizes, main="", xlab="Max Cluster Size")
abline(v = q1, col=2, lty=2); abline(v = obs, col=3, lty=2)
```

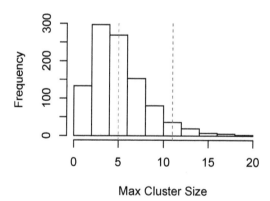

Max Cluster Size

As in the previous section, quantile points (rejection limit points) at a certain number of iterations were randomly extracted and the distribution of quantile points at each number of iterations was illustrated.

```
nreps = seq(10, length(maxcsizes), length=100)
q1s = sapply(nreps, function(x)
quantile(maxcsizes[2:x], 0.95))
par(mfrow=c(1,1))
```

```
plot(nreps, q1s, main="", xlab="Number of permutations",
ylab="95% point", type="b")
```

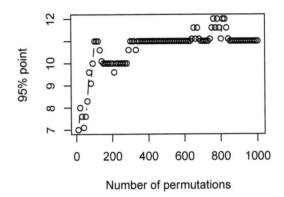

Number of permutations

As a result, the convergence appears to have begun to some extent with the number of iterations. Since it is necessary to perform experiments with various settings, it is better to perform as many iterations as possible, rather than determining a small number of iterations.

Chapter 4

Multivariate Approach

Brain image data is useful for the early detection of diseases and the elucidation of the disease state because in a three-dimensional space, it can anatomically describe the morphology and function of the brain. In the brain image data, the three-dimensional structure of the brain is divided into a large number of cubes of approximately one million voxels arranged in a three-dimensional space and numerical values (voxel values) are recorded in units of voxels. When evaluating multiple voxels simultaneously, it is necessary to take into account the dependence of neighboring voxel values, namely, the spatial dependence within each individual's brain image. This chapter introduces a method for analyzing high-dimensional data with such dependencies.

Data reshape

To analyze the brain image data measured in the three-dimensional space, it is necessary to convert the format such as subjects in rows and variables in columns (Figure 4.1). Therefore, the brain image data of each subject is vectorized. The 3D image data of N voxels for n subjects is in the following $n \times N$ matrix format.

$$X = (x_1, x_2, \ldots, x_N)$$

Since a column corresponds to a voxel position, the column vector is a voxel value for a certain coordinate. It can be viewed as ultra-high-dimensional (big) data with approximately one million columns. In addition, since the combination of multiple columns represents a certain (anatomically meaningful) brain region, the column sum is the ROI (Region of Interest) value for each subject. Thus, the columns of the brain image data matrix correspond to the spatial positions.

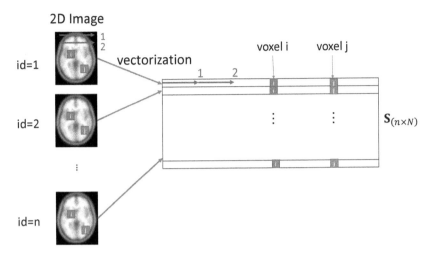

Figure 4.1: Brain image data vectorization and data matrix.

High-dimensional data analysis methods include matrix decomposition methods, sparse estimation methods, machine-learning methods and deep learning methods. First, the basics of principal component analysis were explained as matrix decomposition methods, then it describes the application of this method in brain image analysis as well as the relationship of this method with supervised learning methods and two-steps dimension reduction methods that are useful in brain image analysis. These are explained using the R code.

R example

Preparation

First, we prepare and then we create the data matrix. Install package (as necessary)

```
if(!require("mand")) install.packages("mand")
```

Load package

```
library(mand)
```

Load additional data

```
load(file = "example.RData")
load(file = "atlas.RData")
```

Generate Simulation Data

The mand package has a function to generate simulation brain data from the base image, the difference image and the standard deviation image. These basic images are loaded as follows.

```
data(baseimg)
data(diffimg)
data(mask)
```

The number of voxels in the original 3D image is as follows.

```
dim(baseimg)
```

```
## [1] 30 36 30
```

To understand the result easily, the difference region was restricted to Parahippocampus and Hippocampus.

```
diffimg2 = diffimg * (tmpatlas %in% 37:40)
```

An image data matrix with subjects in the rows and voxels in the columns was generated by using the simbrain function.

```
img1 = simbrain(baseimg = baseimg, diffimg = diffimg2,
sdevimg=sdevimg, mask=mask, n0=20, c1=0.01, sd1=0.05)
```

The base image, the difference image and the standard deviation image were specified in the first three arguments. The out-of-brain region was specified by the mask argument, which was the binary image. The remaining arguments are the number of subjects per group, the coefficient multiplied by the difference image and the standard deviation for the noise.

The data matrix dimension was 40(subject) x 6422(voxel).

```
dim(img1$S)
```

```
## [1]    40 6422
```

The rec function creates the 3D image from the vectorized data (the first subject).

```
coat(rec(img1$S[1,], img1$imagedim, mask=img1$brainpos))
```

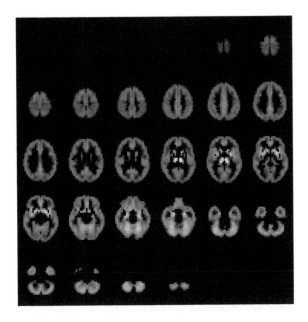

The standard deviation image is created from the resulting data matrix.

```
sdimg = apply(img1$S, 2, sd)
coat(template, rec(sdimg, img1$imagedim, mask=img1$brainpos))
```

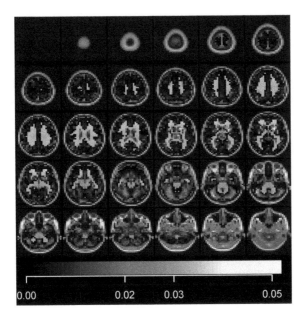

Matrix decomposition

Consider the following matrix decomposition for an $n \times N$ data matrix when $\boldsymbol{X}$ is obtained from an image as in the previous section.

$$\boldsymbol{X} = \boldsymbol{s}\boldsymbol{w}^\top$$

Here, $\boldsymbol{s}$ is a n-dimensional score vector and $\boldsymbol{w}$ is a N-dimensional weight vector. The score vector represents the personal data calculated from the images and analyzes the relationship with clinical data and is useful for predicting and diagnosing the onset (progression). A score expression is a summary or dimension reduction of the data. It is also a latent variable that is not directly observed but is calculated through weights from other observed variables.

The matrix decomposition formula is transformed as follows.

$$\boldsymbol{s} = \boldsymbol{X}\boldsymbol{w}$$

Thus, the weight vector represents the spatial contribution of the image in the score. The weight vector $\boldsymbol{w}$ can be obtained by optimizing the score $\boldsymbol{s}$ from the data. The most typical example is optimization that maximizes the variance of the score, which is called principal component analysis. Additionally, there are other matrix decomposition (score representation) methods such as independent component analysis (ICA), partial least squares (PLS), and non-negative matrix factorization (NMF), but the difference is in the score optimization method.

Principal component analysis

Principal component analysis (PCA) maximizes the variance of the score such that the score can be viewed as a new variable for the axis where the data variability is large. The variance $\mathrm{var}(\boldsymbol{s})$ for the score $\boldsymbol{s} = \boldsymbol{X}\boldsymbol{w}$ is $\mathrm{var}(\boldsymbol{s}) = \boldsymbol{w}^\top \mathrm{var}(\boldsymbol{X})\boldsymbol{w}$. If the data is centered (subtracted by the average from the data for each column), $\mathrm{var}(\boldsymbol{X}) = \boldsymbol{X}^\top \boldsymbol{X}$. However, if $\boldsymbol{w}$ is increased as much as possible, the variance can be increased; thus, the constraint $\boldsymbol{w}^\top \boldsymbol{w} = 1$ is included. Optimization under these constraints can be solved by Lagrange's undetermined multiplier method. The objective function is

$$L(\boldsymbol{w}) = \boldsymbol{w}^\top \boldsymbol{X}^\top \boldsymbol{X}\boldsymbol{w} - \eta \boldsymbol{w}^\top \boldsymbol{w}$$

where $\eta > 0$ is an undetermined multiplier. The weight vector $\boldsymbol{w}$ is obtained by maximizing $L(\boldsymbol{w})$. This requires partial differentiation of $L(\boldsymbol{w})$ with respect to $\boldsymbol{w}$, which is as follows.

$$\frac{\partial L(\boldsymbol{w})}{\partial \boldsymbol{w}} = 2\boldsymbol{X}^\top \boldsymbol{X}\boldsymbol{w} - 2\eta \boldsymbol{w}$$

By solving $\frac{\partial L(\boldsymbol{w})}{\partial \boldsymbol{w}} = 0$, the optimal solution for $\boldsymbol{w}$ is obtained, but you also get the equation $\boldsymbol{X}^\top \boldsymbol{X}\boldsymbol{w} = \eta \boldsymbol{w}$. This is an eigenvalue problem for the matrix $\boldsymbol{X}^\top \boldsymbol{X}$. Thus

w is the eigenvector of the matrix $X^\top X$. As a solution, consider the following optimization problem for a certain n-dimensional vector t.

$$L'(w) = t^\top X w - \eta w^\top w \qquad (4.1)$$

The partial derivative is $\frac{\partial L'(w)}{\partial w} = 2t^\top X - 2\eta w$ and $\frac{\partial L'(w)}{\partial w} = 0$ is solved, $w = (t^\top X)/\eta$, so $\hat{w} = \tilde{w}/\|\tilde{w}\|$ (if $\|w\| > 0$), $\hat{w} = 0$ (otherwise) is considered as a solution for $w^\top w = 1$. This solution depends on a vector t, which is given by the algorithm as the initial score. The score is updated in the form of $s \leftarrow X\hat{w}$ according to $\hat{w}$ obtained here and the update is repeated until the value does not change (converges) even if s is updated. Therefore, this solution $\hat{w}$ is also called a coordinate update.

Once the score is obtained, it is called the first principal component (score) and another score is obtained as the second principal component. It can be obtained by deflating X. In other words, for the predicted value $X = ss^\top(s^\top s)^{-1}X$, the residual $X - \hat{X}$ is used as the new X, and the weight is obtained by the above method. If $s^{(k)}$ is the k-th principal component score thus $s^{(k)}$ and $s^{(\ell)}(k \neq \ell)$ obtained become orthogonal. Repeating this, K principal component scores for a predetermined K are obtained.

R example

If the input is a matrix, a principal component analysis is implemented by the msma function of the msma package. Principal component analysis with the number of components equal to 2 for the image data matrix is performed as follows.

```
(fit111 = msma(img1$S, comp=2))
```

```
## Call:
## msma.default(X = img1$S, comp = 2)
##
## Numbers of non-zeros for X:
##          comp1 comp2
## block1   6422  6422
##
## Numbers of non-zeros for X super:
## comp1 comp2
##    1     1
```

The scatter plots for the score vectors specified by the argument v. The argument axes is specified by the two length vectors on which components are displayed.

```
plot(fit111, v="score", axes = 1:2, plottype="scatter")
```

block

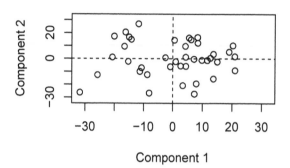

The weight (loading) vectors can be obtained and reconstructed as follows.

```
midx = 1 ## the index for the modality
vidx = 1 ## the index for the component
Q = fit111$wbX[[midx]][,vidx]
outstat1 = rec(Q, img1$imagedim, mask=img1$brainpos)
```

The reconstructed loadings as images are overlayed on the template.

```
coat(template, outstat1)
```

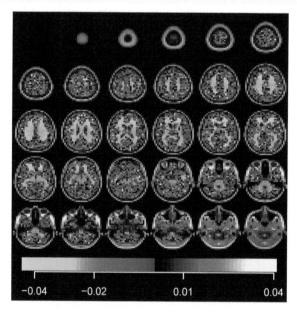

The output is unclear; however, this will be improved later.

Two-steps dimension reduction

To statistically analyze brain image data, we introduce two-steps dimension reduction as an efficient reduction method. In the first stage, the number of dimensions of the image data is reduced by applying base expansion considering the correlation near the voxels. In the second stage, the matrix decomposition method is used taking into account the statistical correlation value calculated from the data values. As introduced in the previous section, there are many possible methods for matrix decomposition at the second stage, but it is useful for brain image analysis to incorporate sparse estimation and supervised learning (Kawaguchi and Yamashita, 2017; Kawaguchi, 2019). The basic expansion method in the first stage is explained.

Basis expansion

X_0 is a $n \times N$ image data matrix and brain images are vectorized in columns. Here, B is an $N \times q$ matrix and the column is a radial B-spline basis function $\phi_\ell(w)$ given as follows.

$$\phi_\ell(w) = \frac{1}{4h^2} \times \begin{cases} \begin{aligned} &h^3 + 3h^2(h - d_\ell(w)) \\ &+ 3h(h - d_\ell(w))^2 - 3(h - d_\ell(w))^3 \end{aligned} & (d_\ell(w) \leq h) \\[2ex] (2h - d_\ell(w))^3 & (h < d_\ell(w) \leq 2h) \\[2ex] 0 & (d_\ell(w) > 2h) \end{cases}$$

$$(4.2)$$

where $d_\ell(w) = \|w - r_\ell\|$ and $r_\ell \in \mathbb{Z}^3 (\ell = 1, 2, \ldots, q)$ is the distance between knots and voxels that are equally spaced at the coordinates on the image, $h > 0$ is the distance between diagonally adjacent knots, and the more distant the voxels are from the knots, the smaller $\phi_\ell(w)$ becomes radially. The ℓ-th column of matrix B corresponds to the ℓ-th knot. Thus, by setting $X = X_0 B$, the number of columns (dimensions) is reduced from N to q. The size of q depends on the knot spacing. For example, when arranged at intervals of 4 voxels, $q \doteqdot 4^{-3} \times N$. When $N = 1,000,000$, $q = 15,625$.

The matrix B is a function that takes the weighted sum of the voxels near the knots at the same time as the dimension reduction and the set of voxels can be thought of as a "region." If the matrix decomposition method is applied to the dimension-reduced X, $S = XW$ is obtained and the score $S = X_0 BW$ and BW can be regarded as the weight for the N-dimensions of the original image. In the first stage of dimension reduction, other basis functions can be considered. Additionally, to augment the basis function, it is also possible to use the sum of the segmented regions using an anatomical brain atlas. Since each atlas has

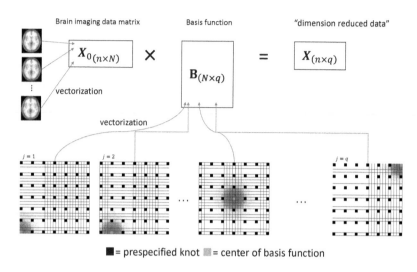

Figure 4.2: Basis expansion.

various independent research outcomes, the results are different depending on which atlas is selected.

Sparse estimation

Thus far, we have been trying to find the weight vector for matrix decomposition by optimizing (maximizing) the objective function. The weight vector found in this way is used to indicate which region contributes to the score so that the corresponding voxel represents the brain region. In this case, it is interpreted as a score relating to a region having a large weight, but if these regions are selected, interpretation becomes easier. For this purpose, often weights smaller than some threshold absolute value are treated as zero. However, Cadima and Jolliffe (1995) show that this can be misleading. Jolliffe et al. (2003) imposed L1 penalty on the maximum variance of principal components, and derived an efficient algorithm called "SCoTLASS" to produce some zero loadings in the components. Zou et al. (2006) proposed an estimation method for sparse principal components based on the regression error property. Witten et al. (2009) and Shen and Huang (2008) presented a penalized matrix decomposition approach to obtain sparse principal component loadings.

For the approach, additional constraints are added, as well as sparse estimation for weights, meaning that small weights can be estimated to be zero. The objective function for sparse principal component analysis is given by

$$L'_\lambda (w) = t^\top X w - \eta w^\top w - P_\lambda (w).$$

This is obtained by subtracting $P_\lambda(w)$ from the equation (4.1). If $P_\lambda(w) = \lambda |w|$, it becomes least absolute shrinkage and selection operator (LASSO) estimation. Besides this, other options are smoothly clipped absolute deviation (SCAD), minimax concave penalty (MCP), trun-cated L_1-penalty, moderately clipped LASSO, sparse ridge, modified log. In the case of LASSO estimation, the solution of its derivative $\frac{\partial L'_\lambda(w)}{\partial w} = 0$ is

$$\widetilde{w} = h_\lambda(t^\top X)$$

where $h_\lambda(y) = sign(y)(|y| - \lambda)_+$. $\lambda > 0$ is a regularization parameter. A weight smaller than λ is 0 and it can be interpreted that a region corresponding to a non-zero weight contributes to the score. This is advantageous not only in interpretation but also in estimation for high-dimensional data. This λ is selected using indicators such as Bayesian information criterion (BIC) and cross validation (CV), as illustrated in the R code later.

We present a lemma provided by Shen and Huang (2008).

Lemma 1 *Let $\hat{\beta}$ be the minimizer of $\beta^2 - 2y\beta + P_\lambda(|\beta|)$. For the penalty $P_\lambda(|\theta|) = 2\lambda|\theta|$, the $\hat{\beta}$ is given by*

$$\hat{\beta} = h_\lambda(y) = sign(y)(|y| - \lambda)_+,$$

where $(x)_+ = \max(0, x)$.

The penalty function $P_\lambda(|\theta|) = 2\lambda|\theta|$ is called the soft thresholding penalty. The lemma in Shen and Huang (2008) also provides the solution for the hard thresholding penalty $P_\lambda(|\theta|) = \lambda^2 I(|\theta| \neq 0)$.

$$\widetilde{w} = I(|y| - \lambda)y$$

The SCAD penalty Fan and Li (2001) is defined by

$$P_{\lambda,\eta}(w) = \begin{cases} \lambda w & (w \leq \lambda) \\ \frac{\eta\lambda w - 0.5(w^2 + \lambda^2)}{\eta - 1} & (\lambda < w \leq \eta\lambda) \\ \frac{\lambda^2(\eta^2 - 1)}{2(\eta - 1)} & (otherwise) \end{cases}$$

The derivative is given as

$$\frac{\partial P_{\lambda,\eta}(w)}{\partial w} = \begin{cases} \lambda & (w \leq \lambda) \\ \frac{\eta\lambda - w}{\eta - 1} & (\lambda < w \leq \eta\lambda) \\ 0 & (otherwise) \end{cases}$$

The solution is given in a lemma in Shen and Huang (2008) as

$$\widetilde{w} = \begin{cases} h_\lambda(y) & (w \leq \lambda) \\ \frac{h_{\eta\lambda/(\eta-1)}(y)}{1 - 1/(\eta - 1)} & (\lambda < w \leq \eta\lambda) \\ y & (otherwise) \end{cases}$$

Zhang et al. (2010) proposed the MCP for $\lambda \geq 0$ and $\eta > 0$.

$$P_{\lambda,\eta}(w) = \begin{cases} \lambda w - \frac{w^2}{2\eta} & (w \leq \eta\lambda) \\ \frac{\eta\lambda^2}{2} & (otherwise) \end{cases}$$

The derivative is given as

$$\frac{\partial P_{\lambda,\eta}(w)}{\partial w} = \begin{cases} \lambda - \frac{w}{\eta} & (w \leq \eta\lambda) \\ 0 & (otherwise) \end{cases}$$

The solution is given as

$$\widetilde{w} = \begin{cases} \frac{h_\lambda(y)}{1 - 1/\eta} & (w \leq \eta\lambda) \\ y & (otherwise) \end{cases}$$

These differences will be shown later in the example with R.

Combining this sparse estimation with the aforementioned two-step dimension reduction, we obtain the composite basis function. The column of matrix $\boldsymbol{W}$ in the matrix $\boldsymbol{BW}$ that we have just come up is sparse. This is illustrated in Figure 4.3.

Each column of matrix $\boldsymbol{B}$ has the radial basis function defined earlier, which has the shape of a circle. Each column of $\boldsymbol{BW}$ is a linear combination of these circular basis functions and has a sparse structure in which only the necessary parts are non-zero to represent the data. This is called the composite basis function and can be a flexible shape as shown in Figure 4.3. Thus, one of the advantages of the two-step dimensionality reduction is that the original form of the basis function can be arbitrary and the analysis can be performed with a flexible form of the basis function.

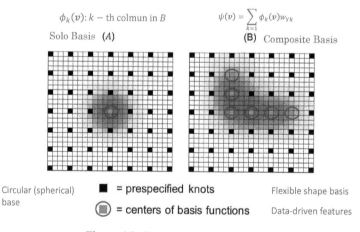

$\phi_k(v): k - $ th colmun in B 　　　 $\psi(v) = \sum_{k=1} \phi_k(v)w_{Yk}$

Solo Basis **(A)** 　　　 **(B)** Composite Basis

Circular (spherical) base 　 ■ = prespecified knots 　 Flexible shape basis

⬤ = centers of basis functions 　 Data-driven features

Figure 4.3: Composite basis expansion.

Supervised estimation

In brain images, weights may be obtained from vast voxel values and scores that maximize variance are not necessarily related to clinical information. To increase the variance from the scores and the correlation between the score as well as the vector **Z** representing certain clinical information, the following optimization problem is considered.

$$L(\mathbf{w}) = (1-\mu)\mathbf{s}^\top \mathbf{s} + \mu \mathbf{s}^\top \mathbf{Z}$$

This **Z** is a supervisor (clinical information), and the score has both axes in the directions of the image data variation and clinical information, while the balance is taken with the parameter μ ($0 \le \mu \le 1$). Depending on the manner in which μ is taken, weights are often higher in areas that are clinically easy to interpret. Based on the results of previous studies, $\mu = 0.5$ is recommended (Kawaguchi, 2019).

R example

Methods

This is an example of the two-step dimension reduction.

Generate radial basis function.

```
B1 = rbfunc(imagedim=img1$imagedim, seppix=2, hispec=FALSE,
mask=img1$brainpos)
```

Multiplying the basis function to image data matrix.

```
SB1 = basisprod(img1$S, B1)
```

The original dimension was 6422.

```
dim(img1$S)
```

```
## [1]    40 6422
```

The dimension was reduced to 813.

```
dim(SB1)
```

```
## [1]    40 813
```

Principal Component Analysis (PCA)

The PCA is applied to the dimension-reduced image.

```
(fit211 = msma(SB1, comp=2))

## Call:
## msma.default(X = SB1, comp = 2)
##
## Numbers of non-zeros for X:
##        comp1 comp2
## block1   813   813
##
## Numbers of non-zeros for X super:
## comp1 comp2
##     1     1
```

The loading is reconstructed to the original space by using the `rec` function.

```
Q = fit211$wbX[[1]][,1]
outstat1 = rec(Q, img1$imagedim, B=B1, mask=img1$brainpos)
```

The plotted (sign-flipping) loading is smoother than the one without the dimension reduction by the basis function.

```
outstat2 = -outstat1
coat(template, outstat2)
```

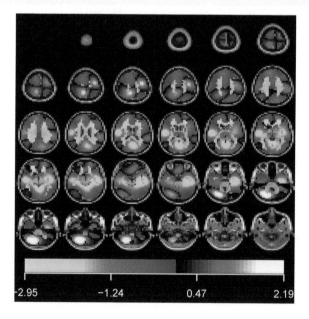

Sparse PCA

If `lambdaX` (>0) is specified, a sparse principal component analysis is implemented.

```
(fit112 = msma(SB1, comp=2, lambdaX=0.075))
```

```
## Call:
## msma.default(X = SB1, comp = 2, lambdaX = 0.075)
##
## Numbers of non-zeros for X:
##        comp1 comp2
## block1    37    28
##
## Numbers of non-zeros for X super:
## comp1 comp2
##     1     1
```

The plotted loading is narrower than that from the PCA.

```
Q = fit112$wbX[[midx]][,vidx]
outstat1 = rec(Q, img1$imagedim, B=B1, mask=img1$brainpos)
outstat2 = outstat1
coat(template, outstat2)
```

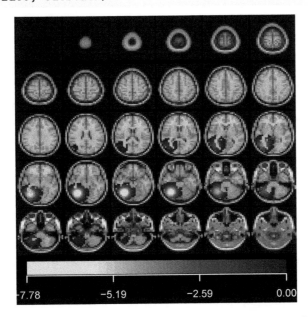

The region is reported as follows to be compared with the next method.

```
atlastable(tmpatlas, outstat2, atlasdataset)
```

```
##      ROIid                   ROIname sizepct sumvalue
## 56     56           Right Fusiform   0.948 -341.276
## 54     54 Right Inferior Occipital   1.000 -115.863
## 98     98        Right Cerebellum 6   1.000 -202.782
## 90     90  Right Inferior Temporal   0.887 -208.641
## 92     92         Right Cerebellum   1.000 -301.399
## 48     48             Right Lingual   1.000 -139.797
## 96     96      Right Cerebellum 4-5   1.000  -73.940
## 86     86      Right Middle Temporal   0.931 -116.613
## 40     40      Right Parahippocampus   0.860  -34.300
## 52     52      Right Middle Occipital   1.000  -33.213
##         Min.    Mean Max.
## 56 -7.779 -0.011   0
## 54 -6.967 -0.004   0
## 98 -6.858 -0.006   0
## 90 -6.634 -0.006   0
## 92 -6.442 -0.009   0
## 48 -4.917 -0.004   0
## 96 -4.708 -0.002   0
## 86 -4.345 -0.004   0
## 40 -3.412 -0.001   0
## 52 -2.675 -0.001   0
```

Supervised Sparse PCA

The simbrain generates the synthetic brain image data and the binary outcome. The outcome Z is obtained.

```
Z = img1$Z
```

If the outcome Z is specified in the msma function, a supervised sparse principal component analysis is implemented.

```
(fit113 = msma(SB1, Z=Z, comp=2, lambdaX=0.075, muX=0.5))
```

```
## Call:
## msma.default(X = SB1, Z = Z, comp = 2, lambdaX = 0.075, muX = 0.5)
##
## Numbers of non-zeros for X:
##         comp1 comp2
## block1    40    38
##
## Numbers of non-zeros for X super:
```

```
## comp1 comp2
##     1     1
```

The plotted loading is located differently from the sparse PCA.

```
Q = fit113$wbX[[1]][,1]
outstat1 = rec(Q, img1$imagedim, B=B1, mask=img1$brainpos)
outstat2 = -outstat1
coat(template, outstat2)
```

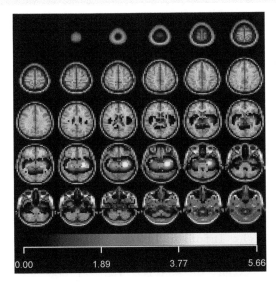

The region near the hippocampus, which differs from the sparse PCA (without supervision).

```
atlastable(tmpatlas, outstat2, atlasdataset)
```

##	ROIid	ROIname	sizepct	sumvalue	Min.
## 37	37	Left Hippocampus	1.000	153.060	0
## 39	39	Left Parahippocampus	1.000	143.699	0
## 55	55	Left Fusiform	0.965	110.647	0
## 38	38	Right Hippocampus	1.000	75.242	0
## 73	73	Left Putamen	1.000	29.517	0
## 41	41	Left Amygdala	1.000	26.821	0
## 95	95	Left Cerebellum 4-5	1.000	42.869	0
## 40	40	Right Parahippocampus	1.000	90.091	0
## 89	89	Left Inferior Temporal	1.000	78.407	0
## 77	77	Left Thalamus	1.000	48.574	0

```
##      Mean  Max.
## 37 0.005 5.660
## 39 0.004 5.306
```

```
## 55 0.003 4.703
## 38 0.002 3.683
## 73 0.001 3.654
## 41 0.001 3.614
## 95 0.001 3.585
## 40 0.003 3.462
## 89 0.002 3.430
## 77 0.001 3.083
```

The loading for the second component

```
Q = fit113$wbX[[1]][,2]
outstat1 = rec(Q, img1$imagedim, B=B1, mask=img1$brainpos)
outstat2 = -outstat1
coat(template, outstat2)

atlastable(tmpatlas, outstat2, atlasdataset)
```

##	ROIid	ROIname	sizepct	sumvalue
## 56	56	Right Fusiform	0.948	-335.918
## 98	98	Right Cerebellum 6	1.000	-207.224
## 54	54	Right Inferior Occipital	1.000	-118.017
## 92	92	Right Cerebellum	1.000	-339.879
## 90	90	Right Inferior Temporal	0.887	-204.087
## 48	48	Right Lingual	1.000	-141.114
## 96	96	Right Cerebellum 4-5	1.000	-67.812
## 86	86	Right Middle Temporal	0.931	-110.785
## 40	40	Right Parahippocampus	0.860	-29.013
## 52	52	Right Middle Occipital	1.000	-32.460

##	Min.	Mean	Max.
## 56	-7.904	-0.010	0
## 98	-7.112	-0.006	0
## 54	-7.083	-0.004	0
## 92	-6.865	-0.010	0
## 90	-6.495	-0.006	0
## 48	-5.258	-0.004	0
## 96	-4.432	-0.002	0
## 86	-4.319	-0.003	0
## 40	-3.012	-0.001	0
## 52	-2.641	-0.001	0

This is similar to the result from the sparse PCA (without supervision).

The following method can be used to plot the weights of several components simultaneously. It is first reconstructed in three dimensions with the multirec function and then plotted with the `multicompplot` function. It is set to display four columns per component.

```
ws = multirec(fit113, imagedim=img1$imagedim, B=B1,
mask=img1$brainpos)
multicompplot(ws, template, col4comp=4)
```

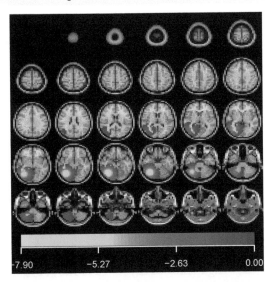

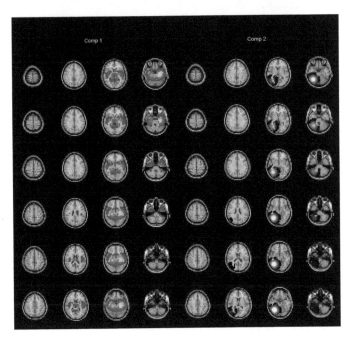

Characteristics

In this section, we perform some experiments to show the characteristics of the methods in the previous section.

Impact seppix

The basis function is parameterized by the radius, which is the argument `seppix` of the `rbfunc` function. The `msma` function is applied.

```
seppixs = 2:7
fit115s = lapply(seppixs, function(sp){
B1 = rbfunc(imagedim=img1$imagedim, seppix=sp,
 hispec=FALSE, mask=img1$brainpos)
SB1 = basisprod(img1$S, B1)
fit=msma(SB1, Z=Z, comp=2, lambdaX=0.075, muX=0.5)
list(fit=fit, B1=B1)
})
```

```
par(mfrow=c(2,3), mar=c(1,2,1,2))
for(i in 1:length(seppixs)){
Q = fit115s[[i]]$fit$wbX[[midx]][,vidx]
outstat1 = rec(Q, img1$imagedim, B=fit115s[[i]]$B1,
mask=img1$brainpos)
coat(template, -outstat1, pseq=10,color.bar=FALSE,
paron=FALSE, main=paste("seppix =", seppixs[i]))
}
```

With the larger value it is difficult to determine the region.

Impact lambda

The iterative fits with different regularized parameters are implemented specifying them using the argument `lambdaX`.

```
lambdaXs = round(seq(0, 0.2, by=0.005), 3)
fit114s = lapply(lambdaXs, function(lam)
msma(SB1, Z=Z, comp=2, lambdaX=lam, muX=0.5, type="lasso") )
```

For some of the parameters, the sparsity can be controlled by λ, as shown below.

```
lambdaXs2 = c(0, 0.025, 0.05, 0.075, 0.1, 0.15)
par(mfrow=c(2,3), mar=c(1,2,1,2))
for(i in which(lambdaXs %in% lambdaXs2)){
Q = fit114s[[i]]$wbX[[1]][,1]
```

```
outstat1 = rec(Q, img1$imagedim, B=B1, mask=img1$brainpos)
coat(template, -outstat1, pseq=10,color.bar=FALSE,
 paron=FALSE, main=paste("lambda =", lambdaXs[i]))
}
```

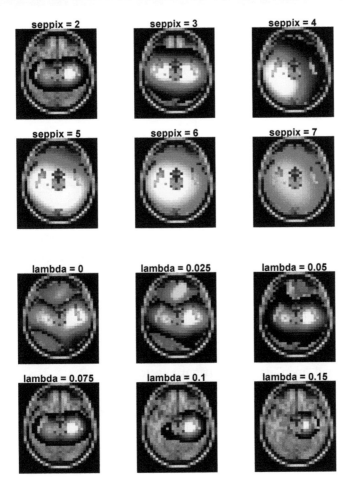

The optimal value can be determined by the BIC.

```
nzwbXs = unlist(lapply(fit114s, function(x) x$nzwbX[2]))
BICs = unlist(lapply(fit114s, function(x) x$bic[2]))
(optlam = lambdaXs[which.min(BICs)])
```

```
## [1] 0.085
```

```
(optnzw = nzwbXs[which.min(BICs)])
```

```
## [1] 27
```

The plot of BIC values against the number of non-zero loadings and the λ.

```
par(mfrow=c(1,2))
plot(lambdaXs, BICs, ylab="BIC")
abline(v=optlam, col="red", lty=2)
plot(nzwbXs, BICs, ylab="", log="x")
abline(v=optnzw, col="red", lty=2)
```

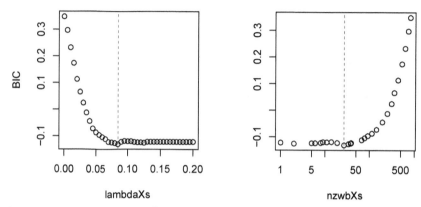

The optimal BIC value was $\lambda = 0.085$ which was relatively better among reconstructed loading images with different λ's as shown above.

Penalty Functions

For the `msma` function, four types of penalty functions are currently available. These depend on parameters, "lasso" and "hard" which depend on one parameter, and "scad" and "mcp" depend on two parameters.

```
penalties2 = c("lasso", "hard", "scad", "mcp")
etas = list(lasso=1, hard=1, scad=c(1, 3.7), mcp=c(2, 3))
```

Using the internal function `sparse` in the `mand` package, we see how each penalty function performs the transformation.

```
xs = seq(-6, 6, by=0.1)
par(mfrow=c(2,3), mar=c(2,2,3,2))
for(p1 in penalties2){
etal = etas[[p1]]
for(e1 in etal){
sout1 = sparse(xs, 2, type=p1, eta=e1)
plot(xs, sout1, xlab="", ylab="",
main=paste(p1, "(eta =", e1, ")"), type="b")
}}
```

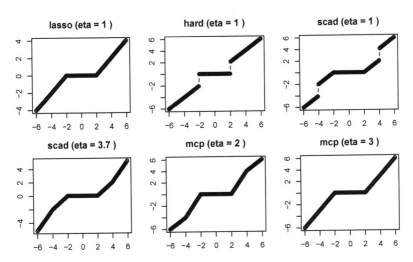

Next, we apply each penalty function in the `msma` function to the brain image data.

```
par(mfrow=c(2,3), mar=c(1,2,1,2))
for(p1 in penalties2){
eta1 = etas[[p1]]
for(e1 in eta1){
fit = msma(SB1, Z=Z, comp=2, lambdaX=0.025, muX=0.5,
 type=p1, eta=e1)
Q = fit$wbX[[midx]][,vidx]
outstat1 = rec(Q, img1$imagedim, B=B1, mask=img1$brainpos)
outstat2 = -outstat1
coat(template, outstat2, pseq=10, color.bar=FALSE,
 paron=FALSE, main=paste(p1, "(eta =", e1, ")"))
}}
```

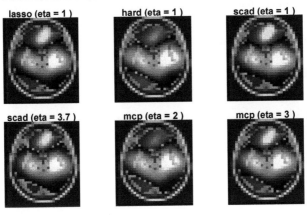

Parameter Selection

The number of components was set to be 30.

```
fit114 = msma(SB1, Z=Z, comp=30, muX=0.5)
```

The cumulative percentage of the explained variance (CPEV) is plotted against the number of components. When the argument v is specified as "cpev", the CPEVs are plotted.

```
plot(fit114, v="cpev")
abline(h=0.8, lty=2)
```

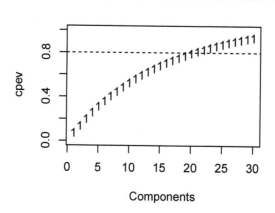

The function `ncompsearch` searches for the optimal value based on the criterion.

```
(ncomp1 = ncompsearch(SB1, Z=Z, muX=0.5,
comps = 50, criterion="BIC"))
```

```
## Optimal number of components: 39 ( BIC )
```

The criterion can use not only BIC but also CV (Cross Validation).

```
(ncomp2 = ncompsearch(SB1, Z=Z, muX=0.5,
comps = c(1, seq(5, 30, by=5)), criterion="CV"))
```

```
## Optimal number of components: 10 ( CV )
par(mfrow=c(1,2))
plot(ncomp1)
plot(ncomp2)
```

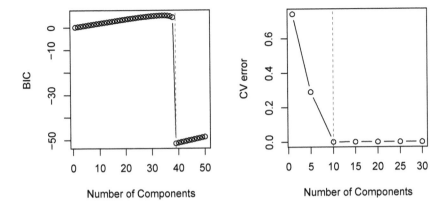

For each component number from 1 to 5, the optimal λ was selected, and the number of non-zero loadings in each component was counted.

```
maxncomp = 5
opts = sapply(1:maxncomp, function(c1){
opt=regparasearch(SB1, Z=Z, comp=c1, muX=0.5)$optlambdaX
fit = msma(SB1, Z=Z, comp=c1, lambdaX=opt, muX=0.5)
nz = rep(NA, maxncomp)
nz[1:c1] = fit$nzwbX
c(c1, round(opt,3), nz)
})

opts1=t(opts)
colnames(opts1) = c("#comp", "lambda",
paste("comp",1:maxncomp))
kable(opts1, "latex", booktabs = T)
```

#comp	lambda	comp 1	comp 2	comp 3	comp 4	comp 5
1	0.031	267				
2	0.056	79	92			
3	0.062	59	69	53		
4	0.056	79	92	85	69	
5	0.075	40	38	48	25	25

From this result, it can be seen that when the number of components is large, a small non-zero loading is selected for each component, and when the number of components is small, a large non-zero loading is selected for each component.

```
(ncomp3 = ncompsearch(SB1, Z=Z, muX=0.5,
lambdaX=0.075, comps = 30, criterion="BIC"))
```

```
## Optimal number of components: 30 ( BIC )
plot(ncomp3)
```

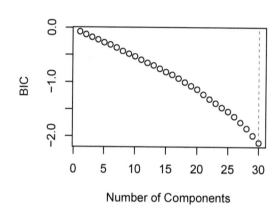

Thus, there are several strategies for selecting λ and the number of components. The msma package has an optparasearch function and the search.method argument allows you to select one of four methods.

The regparaonly method searches for the regularized parameters with a fixed number of components (set to be 5 and 10 as the default).

```
(opt11 = optparasearch(SB1, Z=Z, muX=0.5, comp=5,
search.method = "regparaonly", criterion="BIC"))
```

```
## Search method: regparaonlySearch criterion: BIC
##
## Optimal number of components: 5
##
## Optimal parameters:
##
## [1] 0.087
```

The optimized parameters are used in the msma function as follows.

```
(fit311 = msma(SB1, Z=Z, muX=0.5,
comp=opt11$optncomp, lambdaX=opt11$optlambdaX))
```

```
## Call:
## msma.default(X = SB1, Z = Z, comp = opt11$optncomp,
    lambdaX = opt11$optlambdaX,
##      muX = 0.5)
##
## Numbers of non-zeros for X:
```

```
##         comp1 comp2 comp3 comp4 comp5
## block1    25    26    22    16    14
##
## Numbers of non-zeros for X super:
## comp1 comp2 comp3 comp4 comp5
##    1     1     1     1     1
```

The `regpara1st` identifies the regularized parameters by fixing the number of components, then searching for the number of components with the selected regularized parameters. Note that the default is to search up to 10 components.

```
(opt12 = optparasearch(SB1, Z=Z, muX=0.5,
search.method = "regpara1st", criterion="BIC"))
```

```
## Search method: regpara1stSearch criterion: BIC
##
## Optimal number of components: 10
##
## Optimal parameters:
##
## [1] 0.081
```

```
fit312 = msma(SB1, Z=Z, muX=0.5,
comp=opt12$optncomp, lambdaX=opt12$optlambdaX)
```

The `ncomp1st` method identifies the number of components with a regularized parameter of 0, then searches for the regularized parameters with the selected number of components.

```
(opt13 = optparasearch(SB1, Z=Z, muX=0.5,
search.method = "ncomp1st", criterion="BIC"))
```

```
## Search method: ncomp1stSearch criterion: BIC
##
## Optimal number of components: 1
##
## Optimal parameters:
##
## [1] 0.087
```

```
fit313 = msma(SB1, Z=Z, muX=0.5,
comp=opt13$optncomp, lambdaX=opt13$optlambdaX)
```

If the argument `criterion4ncomp` is specified as `criterion4ncomp="CV"`, only the criteria for selecting the number of components can be changed to CV.

The `simultaneous` method identifies the number of components by searching the regularized parameters in each component.

```
(opt14 = optparasearch(SB1, Z=Z, muX=0.5,
search.method = "simultaneous", criterion="BIC"))

## Search method: simultaneousSearch criterion: BIC
##
## Optimal number of components: 10
##
## Optimal parameters:
##
## [1] 0.081

fit314 = msma(SB1, Z=Z, muX=0.5,
comp=opt14$optncomp, lambdaX=opt14$optlambdaX)
```

In this example, the results were the same except for the `ncomp1st`.

Other methods

Non-negative matrix factorization (NMF) also considers $X = SW$ matrix decomposition for the $n \times N$ matrix X. However, S is $n \times K$, W is $K \times N$ and each element is non-negative. The objective function for $N = 1$ is given as follows.

$$F(w) = \sum_{i=1}^{n} \left(x_i - \sum_{j=1}^{K} s_{ij} w_j \right)^2 = \|X - SW\|_F^2$$

where $W = (w_1, \ldots, w_K)$ is a non-negative weight vector and $\| \ \|_F$ is the frobenium norm. The auxiliary function method can be used to optimize this. By optimizing with the auxiliary function, the objective function can be optimized. The auxiliary function of the function $b(w)$ is $a(w, w')$ that satisfies the following.

$$a(w, w') \geq b(w'), \quad a(w, w) = b(w)$$

In fact, auxiliary function of the $F(w)/2$ is

$$G(w, w') = F(w') + (w - w')\nabla F(w') + 1/2(w - w')^\top K(w')(w - w')$$

where $K(w) = (s^\top s w)_i / w_i$. Here w' is also called an auxiliary variable. A solution that minimizes $F(w)$ can be obtained by iteratively updating the auxiliary function alternately with w and w'. The update is given as follows.

$$w'_{ij} \leftarrow w_{ij} \frac{(X^\top s)_{ij}}{(ws^\top s)_{ij}}, \quad s'_{ij} \leftarrow s_{ij} \frac{(X^\top w)_{ij}}{(sw^\top w)_{ij}}$$

NMF is easy to interpret because all weights are positive. In addition, since NMF is related to k-means clustering, it is often applied to cluster analysis.

Independent component analysis (ICA) can also be seen as matrix decomposition $X = SW$ for X. Suppose there are two scores (sources) s_1 and s_2. The weights are determined so that they are statistically independent. The Infomax algorithm is often used in brain image analysis. Here, we introduce a method based on the maximum likelihood method.

Statistical independence means that the joint density function is the product of its marginal densities. Therefore, when the joint density function $f(s_1, s_2)$ and the density function of s_k is $f(s_k; \theta_k)$, $f(s_1, s_2) = f(s_1; \theta_1) \times f(s_2; \theta_2)$. Thus, the method of calculating the weight is maximizing the log-likelihood function $\ell(W, \theta)$ with the observation $S = XW^{-1}$.

$$\ell(W, \theta) = \sum_{i=1}^{n} \sum_{k=1}^{K} \log f(s_{ki}; \theta_k)$$

The density function parameters and weights are found to be maximized. As for applications in brain image analysis, in functional MRI analysis, it is considered that regions with high weights form components in the brain among components that are extracted, denoised, and spatially independent. It is used for data-driven network estimation. The ICA for fMRI analysis will be introduced in the next chapter.

R example

The NMF can be implemented by invoking the nmf function of the NMF library.

```
if(!require("NMF")) install.packages("NMF")
library(NMF)
```

NMF can be performed on a brain image data matrix with the number of components as 2 in the following way.

```
res = nmf(SB1, 2)
```

Using the coef function, we can extract the weights and plot them on the brain image using the coat function in the same way as when applying the msma function.

```
Q = t(coef(res))[,1]
outstat1 = rec(Q, img1$imagedim, B=B1, mask=img1$brainpos)
coat(template, outstat1)
```

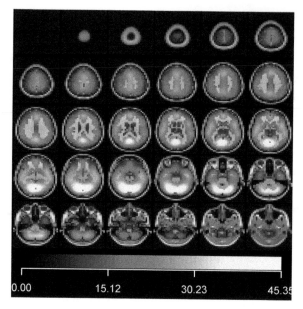

The ICA can be implemented by invoking the `icaimax` function of the `ica` library.

```
if(!require("ica")) install.packages("ica")
library(ica)
```

ICA using the information-maximization (Infomax) approach can be performed on a brain image data matrix with the number of components as 2 in the following way.

```
imod = icaimax(SB1,2)
```

In this case, the weights are extracted as follows and plotted on the brain image in the same way.

```
Q = imod$M[,1]
outstat1 = rec(Q, img1$imagedim, B=B1, mask=img1$brainpos)
coat(template, outstat1)
```

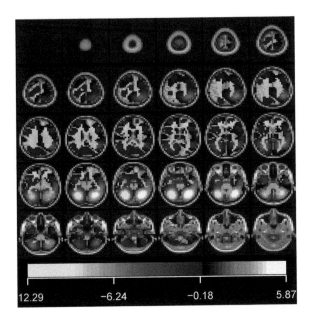

| 12.29 | −6.24 | −0.18 | 5.87 |

Cluster analysis

Clustering refers to classifying objects that have not been classified into several groups (clusters) consisting of similar objects according to the similarity between data. In this case, there is no correct answer (teacher or supervisor) to be predicted or judged and it is classified as "unsupervised learning." The clusters must be given meaning to be classified according to the features represented by the data. In brain image analysis, it is also used for disease subgrouping based on image data and segmentation during preprocessing.

Approaches to clustering are broadly divided into hierarchical and non-hierarchical methods. The first hierarchical method is to group the most similar combinations in order. Given data of n objects, an initial state is first created with n clusters containing only one object. The distance $d(A,B)$ between the two clusters for all pairs is computed and the two closest clusters are merged. The merging is repeated sequentially until all n objects are merged into one cluster, a hierarchical structure is obtained and finally, a tree diagram (dendrogram) can be created. In other words, the dendrogram illustrates the process of cluster formation, where each terminal node represents each object and the resulting cluster is a binary tree that represents the merged cluster as a non-terminal node. The horizontal axis of the non-terminal nodes represents the distance between clusters when they are merged. Additionally, the dendrograms can be divided into several groups by cutting them at appropriate distances according to the subjective opinion of each analyst.

There are various methods for calculating the "distance" between objects and clusters. The Euclidean distance (Pythagorean theorem for a plane) is the most common and the distance between $x = (x_1, x_2, ..., x_p)^\top$ and $y = (y_1, y_2, ..., y_p)^\top$ can be expressed by the following formula.

$$d(x, y) = \sqrt{(x_1 - y_1)^2 + (x_2 - y_2)^2 + \cdots + (x_p - y_p)^2}$$

There are other Mahalanobis distances, Chebyshev distances, Minkowski distances, but please refer to the literature (e.g. Aggarwal and Reddy, 2014) for further information.

The following methods depend on the difference of the distance function $d(A, B)$ between clusters A and B.

1. Ward method

   Suppose there are n_A members in cluster C_A. Let the observations of each member be $(x_{i1}^A, x_{i2}^A, ..., x_{ip}^A)$. The sum of squared deviations within a cluster is given by

   $$SS_A = \sum_{i \in A} \sum_{j \in B} (x_{ij}^A - \bar{x}_j^A)^2$$

   where $x_j^A = \sum_{i \in A} x_{ij}^A / n_A$. In this case, the distance between clusters is defined as follows.

   $$d_{AB} = SS_{AB} - (SS_A + SS_B)$$

   where SS_{AB} is the sum of squared deviations when clusters A and B are combined.

2. Single linkage

   $$d_{AB} = \min_{i \in A, j \in B} d_{ij}$$

3. Complete linkage

   $$d_{AB} = \max_{i \in A, j \in B} d_{ij}$$

4. Average linkage

   $$d_{AB} = \frac{1}{n_A n_B} \sum_{i \in A} \sum_{j \in B} d_{ij}$$

The Ward method is highly rated for practicality. Since the Ward method has high classification sensitivity, the chain effect is less likely to occur when creating a dendrogram. The chain effect is a phenomenon in which clusters are formed through a cluster analysis while objects are sequentially and individually absorbed into a specific cluster. This is likely to occur with the shortest distance method. Because of this undesirable phenomenon, the Ward method, in which the chain effect is less likely to occur, is often used.

The non-hierarchical method determines the cluster once the number of clusters is determined. A typical method is the k-means method, which divides n individuals into k clusters by minimizing a particular criterion. The "particular criterion" is the variability within a cluster (sum of squares within a group). There are experimental methods to determine the optimal number of clusters.

The algorithm of the k-means method is as follows.

(1) Initial division: Individuals are divided into k groups. For this purpose, for example, hierarchical clustering is used.

(2) Calculate changes in the criteria for clustering by transferring each individual to another cluster.

(3) Make cluster changes to maximize the criteria for clustering.

(4) Repeat steps (2) and (3) until no further improvement of the clustering standard occurs.

k-means method is known to be equivalent to the NMF (Ding et al., 2005).

Additionally, there is a model-based clustering method, which represents a given data set by the superposition of multiple normal distributions (Gaussian mixture model, GMM). The main feature is that the probability of belonging to each cluster is obtained for each sample and the number of clusters can be determined using the information criterion. It is also used for segmentation in preprocessing brain images. Furthermore, flexible modeling using nonparametric density estimations other than normal distribution is also possible, as described in the next chapter.

When dealing with objects in a high-dimensional space, problems arising from their high-dimensionality are called "curses of dimension." This means that the distance from the center object to other objects increases rapidly with increasing dimensionality. That is, no two objects are similar to each other. Since clustering is an operation to combine similar objects, it is impossible to obtain clusters. A drastic solution to this "curse" is to reduce the number of dimensions by eliminating variables that are deemed unnecessary.

One of the solutions to this problem is the subspace clustering method, which simultaneously performs feature selection, dimension reduction and learning. There are algorithms called CluStream, ORCLUS, and CLIQUE, which are detailed in Rodriguez et al. (2019). In this book, the method combining the dimension reduction method introduced in the previous section and the hierarchical clustering is introduced by using the R code.

R example

The package for drawing more informative dendrograms is loaded.

```
library(dendextend)
```

By inputting fit111, which is the result of PCA being applied to the hcmsma function by the msma function, hierarchical clustering with the score as input is executed.

```
hcmsma111 = hcmsma(fit111)
```

The next step is to plot and draw the dendrogram.

```
dend = as.dendrogram(hcmsma111$hcout)
d1 = color_branches(dend, k=4, groupLabels=TRUE)
labels_colors(d1) = Z[as.numeric(labels(d1))]+1
plot(d1)
```

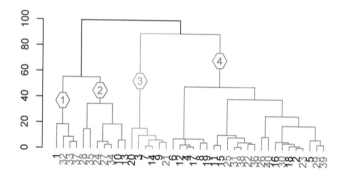

The data was set to be divided into four clusters. The number at the bottom of the dendrogram is the subject number, where black is the control and red is the case.

It seems that clusters 1 and 2 have many cases, and cluster 4 also seems to have many cases. The number is summarized as the matrix with the case or control in the row and the cluster to belong in the column.

```
clus=cutree(d1, 4, order_clusters_as_data = FALSE)
clus=clus[as.character(1:length(clus))]
table(Z, clus)
```

```
##      clus
## Z     1  2  3  4
##    0  1  2  5 12
##    1  3  5  1 11
```

As can be seen, the PCA score did not yield a cluster that completely bisected case and control.

Similarly, we performed clustering using the scores calculated from the sparse PCA.

```
hcmsma112 = hcmsma(fit112)
dend = as.dendrogram(hcmsma112$hcout)
d1 = color_branches(dend, k=4, groupLabels=TRUE)
labels_colors(d1) = Z[as.numeric(labels(d1))]+1
plot(d1)
```

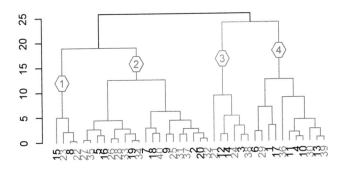

```
clus=cutree(d1, 4, order_clusters_as_data = FALSE)
clus=clus[as.character(1:length(clus))]
table(Z, clus)
```

```
##      clus
## Z    1  2  3  4
##   0  2  8  3  7
##   1  2 11  3  4
```

This result shows that case and control are more sparsely clustered.

Finally, we performed clustering with scores from the supervised sparse PCA.

```
hcmsma113 = hcmsma(fit113)
dend = as.dendrogram(hcmsma113$hcout)
d1 = color_branches(dend, k=4, groupLabels=TRUE)
labels_colors(d1) = Z[as.numeric(labels(d1))]+1
plot(d1)
```

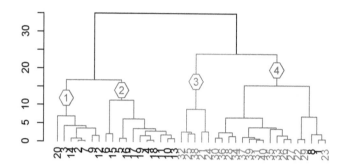

```
clus=cutree(d1, 4, order_clusters_as_data = FALSE)
clus=clus[as.character(1:length(clus))]
table(Z, clus)
```

```
##      clus
## Z    1  2  3  4
##      0  7 11  0  2
##      1  0  0  5 15
```

Because they are supervised by this case or control, the case and control are nicely clustered. This analysis indicates that the case is further divided into two clusters. For practical purposes, it may be useful for disease subtype classification.

Classification method and prediction model

Diagnosis and early detection (prediction) of disease are of interest to many researchers. We consider the presence of disease as an objective, and its posteriori probability can be estimated from image data and can inform a diagnosis, that is, it is a discrimination problem. The objective variable (binary) and the explanatory variable are opposite to GLM.

In discriminant analysis, a data set (y, x) is prepared. Here, y is a group variable (0 = healthy group, 1 = disease group) and $x = (x_1, x_2, \ldots, x_p)^\top$: explanatory variable (image data). Although there is also the problem of classifying them into three or more groups, we will consider two groups. Since we use the group variable as a teacher and are also part of supervised learning. As shown in Figure 4.4, the data that creates the "rules" for discrimination is called training data. The test data used to evaluate the performance of the discriminant rule is excluded when creating the discriminant rule but is created using training data. Discrimination rules need to be applicable to (predicted for) new patients' data.

Flow of discriminant analysis

1. Create a discriminant rule that can be expressed only by x from n training data using a pair of x and y of the training data

2. Discriminate each individual by applying x of the test data to the discrimination rule (x of the training data may be used)

3. Evaluate the discrimination result (matching the discrimination result with y of the test data)

Note that, as an initial evaluation, performance may also be evaluated on training data.

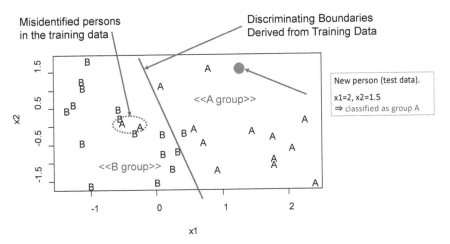

Figure 4.4: Discriminant analysis.

Logistic discrimination

Although there are various classification methods, logistic regression, which directly regresses the classification probability, is useful and extensible. In general, the logistic regression model with the objective variable $Y = 1$ (group A) and $Y = 0$ (group B) and the explanatory variable of x

$$\text{logit}\left(\Pr(Y = 1|\boldsymbol{x})\right) = \boldsymbol{x}^{\top}\boldsymbol{\beta} \qquad (4.3)$$

where $\boldsymbol{\beta} = (\beta_0, \beta_1, \beta_2, \ldots, \beta_p)^{\top}$, $\boldsymbol{x} = (1, x_1, x_2, \ldots, x_p)^{\top}$ and $\text{logit}(p) = \log\left(\frac{p}{1-p}\right)$. The model expression (4.3) can be rewritten as follows.

$$\Pr(Y = 1|\boldsymbol{x}) = \frac{1}{1 + \exp\left(-\boldsymbol{x}^{\top}\boldsymbol{\beta}\right)} \qquad (4.4)$$

Given n subjects data (y_i, x_i) $(i = 1, 2, \ldots, n)$, the likelihood function is

$$L(\boldsymbol{\beta}) = Pr(Y_1 = y_1, \ldots, Y_n = y_n) = \prod_{i=1}^{n} \pi(x_i)^{y_i}(1 - \pi(x_i))^{1-y_i}$$

$\boldsymbol{\beta}$ is required to maximize $L(\boldsymbol{\beta})$ where $\pi(x) = P(Y = 1|x)$. This is equivalent to maximizing the log likelihood $\ell(\boldsymbol{\beta})$ given by

$$\ell(\boldsymbol{\beta}) = \sum_{i=1}^{n} \{y_i \log(\pi(x_i)) + (1 - y_i) \log(1 - \pi(x_i))\} \qquad (4.5)$$

Logistic discrimination is a method for discriminating individuals with $x = x^*$ based on the logistic regression model (4.3). Substitute x^* for x in the probability equation (4.4) we use the rule

$$Pr(Y = 1|x^*) \geq 0.5 \Longrightarrow \text{Belongs to group A}$$
$$Pr(Y = 1|x^*) < 0.5 \Longrightarrow \text{Belongs to group B}$$

The probability $Pr(Y = 1|x)$ is called the diagnostic probability. In actual clinical practice, it is useful to report not only the classification results but also the classification probability.

Dimension reduction

Research on the early diagnosis of dementia using structural brain images has been actively conducted. In Japan, a voxel-based specific regional analysis system for AD (VSRAD) has been developed as a diagnostic tool for AD that is widely used in clinical practice. For VSRAD, diagnosis is made as suspected AD if the volume of the hippocampal region deviates considerably from that of a healthy individual. Such practice constitutes discriminant analysis based on brain images and the region pre-specification also facilitates the interpretation of the results. On the other hand, the identification of an effective region for the discrimination from whole brain is a statistical challenge. The one approach is a statistical multiple (linear) regression analysis, however, it is difficult to estimate the regression parameter because one million voxels compose the high dimensional structure of the variable, which corresponds to approximately one hundred sample sizes.

Imposing some dimension reduction devices in classical multvariate analysis, such as principal component analysis (PCA) or partial least squares (PLS) are also considered in modeling neuroimage data. For such an approach, the two-steps dimension reduction method introduced in the previous section is useful. Reiss and Ogden (2010a) used a two-steps approach consisting of basis expansions and principal component analysis, and studied it theoretically. They also

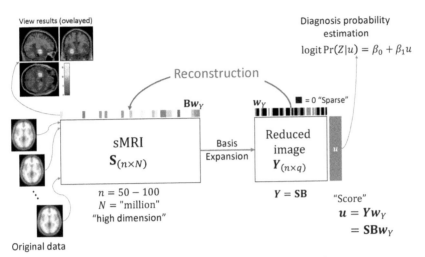

Figure 4.5: Prediction model with dimension reduction.

used the L2 penalty with automatic selection for the regularization parameter based on the method of Wood (2011), which is faster than the grid search used in Casanova et al. (2012) and is independent of the setting for candidates of several smoothing parameter values. As shown in Figure 4.5, the score obtained by applying the sparse component method after dimensionally reducing the original brain image by the basis function is used as an explanatory variable in the logistic regression model to calculate the diagnostic probability.

Empirically, about 10 components may be sufficient for prediction accuracy, so a reasonable estimate can be made even with about 100 cases. Another advantage of this approach is that since it uses matrix operations, weights can be restored to the data dimension and regions important for diagnosis can be displayed in the brain image. Araki et al. (2013) provided a method to select the optimal number of principal components and they classified Alzheimer patients in an early stage based on three-dimensional MRI brain scans. The method is based on a functional logistic regression analysis in Reiss and Ogden (2010a) and is described below.

Suppose we have n independent subjects $\{(y_\alpha, \boldsymbol{x}_\alpha, \boldsymbol{s}_\alpha); \alpha = 1, \ldots, n\}$, where $y_\alpha \in \{0, 1\}$ are binary random variables denoting disease status, $\boldsymbol{x}_\alpha \in \mathbb{R}^p$ are p-dimensional covariate vectors and $\boldsymbol{s}_\alpha = (s_\alpha(\boldsymbol{v}_1), \ldots, s_\alpha(\boldsymbol{v}_N))^\top$ are N-dimensional vectors of image data for the α-th subject defined at voxel location $\boldsymbol{v}_j \in \mathbb{Z}^3 (j = 1, \ldots, N)$. We consider the following logistic model based on covariates and image data,

$$\text{logit} \Pr(Y_\alpha = 1 | \boldsymbol{x}_\alpha, \boldsymbol{s}_\alpha) = \sum_{i=1}^{p} x_{\alpha i} \beta_i + \sum_{j=1}^{N} s_\alpha(\boldsymbol{v}_j) f(\boldsymbol{v}_j), \tag{4.6}$$

where $\{\beta_1,\ldots,\beta_p\}$ and $\{f(\boldsymbol{v}_1),\ldots,f(\boldsymbol{v}_N)\}$ are the unknown parameter vectors. The number of voxels N, which expresses the entire brain image, is about a million, while in general the sample size n is less than one hundred ($n << N$). Therefore the estimation of the unknown coefficients $\boldsymbol{f} = (f(\boldsymbol{v}_1),\ldots,f(\boldsymbol{v}_N))^\top$ becomes unstable. This is a typical case of high-dimensional and low-sample size problem, so it needs some forms of regularization or dimension reduction techniques.

We define the coefficients $f(\boldsymbol{v}_j)$ in equation (4.6) as follows.

$$f(\boldsymbol{v}_j) = \sum_{k=1}^{K} \left\{ \boldsymbol{\phi}(\boldsymbol{v}_j)^\top \boldsymbol{w}_k \right\} \gamma_k = \sum_{m=1}^{K} \psi_k \gamma_k, \tag{4.7}$$

where $\psi_k = \boldsymbol{\phi}^\top(\boldsymbol{v}_j)\boldsymbol{w}_k$ are the composite basis functions, $\boldsymbol{w}_k$ are $q(<< N)$ dimensional coefficient vectors, which was explained in the previous section, and the ℓ-th basis function $\phi_\ell(\boldsymbol{v}_j), (\ell = 1,\ldots,q)$ of

$$\boldsymbol{\phi}(\boldsymbol{v}_j) = (\phi_1(\boldsymbol{v}_j),\ldots,\phi_q(\boldsymbol{v}_j))^\top (j = 1,2,\ldots,N) \tag{4.8}$$

is a radial B-spline basis function, expressed as in the equation (4.2).

Let $\boldsymbol{B}$ be a $N \times q$ matrix $\boldsymbol{B} = (\boldsymbol{\phi}(\boldsymbol{v}_1),\ldots,\boldsymbol{\phi}(\boldsymbol{v}_N))^\top$. Then from equations (4.6) and (4.7), the final model will be

$$\text{logitPr}(Y_\alpha = 1 | \boldsymbol{x}_\alpha, \boldsymbol{s}_\alpha) = \boldsymbol{x}_\alpha^\top \boldsymbol{\beta} + \boldsymbol{s}_\alpha^\top \boldsymbol{BW}\boldsymbol{\gamma}. \tag{4.9}$$

where $\boldsymbol{W} = (\boldsymbol{w}_1,\ldots,\boldsymbol{w}_K)$. The unknown parameters $\boldsymbol{\beta} = (\beta_1,\ldots,\beta_p)^\top$ and $\boldsymbol{\gamma} = (\gamma_1,\ldots,\gamma_K)^\top$ are estimated by maximizing the log likelihood function as in the equation (4.5).

The regularization approaches for the model (4.6) such as ridge regression, lasso, and elastic net are applicable to high dimensional data analysis. Araki and Kawaguchi (2019) provide a method to detect the prognostic brain region and to classify Alzheimer patients in an early stage based on three-dimensional MRI brain scans, via L1-type regularized logistic discrimination with the composite basis function expansions. Estimation of $\boldsymbol{\beta}$ and $\boldsymbol{\gamma}$ is as follows. Suppose $A_\alpha(\boldsymbol{\beta},\boldsymbol{\gamma}) = \boldsymbol{x}_\alpha^\top \boldsymbol{\beta} + \boldsymbol{s}_\alpha^\top \boldsymbol{BW}\boldsymbol{\gamma}$ and y_α follows Bernoulli distribution with mean π_α, where $\pi_\alpha = Pr(Y_\alpha = 1 | \boldsymbol{x}_\alpha, \boldsymbol{s}_\alpha)$. In order to enable simultaneous dimensionality reduction and voxel selection, we estimate the parameter vectors $\boldsymbol{\beta}$ and $\boldsymbol{\gamma}$ by the regularization method which maximized the following function

$$l_{\lambda_2}(\boldsymbol{\beta},\boldsymbol{\gamma}) = \sum_{\alpha=1}^{n} (y_\alpha A_\alpha(\boldsymbol{\beta},\boldsymbol{\gamma}) - \log[1 + \exp\{A_\alpha(\boldsymbol{\beta},\boldsymbol{\gamma})\}]) - n\lambda_2 \sum_{k=1}^{K} \eta_k |\gamma_k|, \tag{4.10}$$

where η_k represent predetermined weights and λ_2 is a second regularization patameter which controls the number of 0's in $\boldsymbol{\gamma}$. By imposing such penalty, a few basis functions among K composite basis functions $\boldsymbol{\phi}^\top(\boldsymbol{v}_j)V_k, (k = 1,\ldots,K)$ are used for classification.

Machine learning

Historically, brain image analysis based on tests was often used, but in recent years interest in machine learning (ML) is increasing, which aims to develop algorithms to predict new information by discovering trends and patterns in data obtained by observation. Research and development in the field of artificial intelligence is flourishing, and ML is also included whose fundamental precept is statistical multivariate analysis. Therefore, ML can identify correlations between voxels and inform individual patient diagnosis and prognosis predictions to perform different studies compared to the previously used test-based analyses (Arbabshirani et al., 2017). ML can be divided broadly into two categories: supervised learning and unsupervised learning. In supervised learning, we try to estimate a function for a given classification variable (disease group and healthy group) or numerical value (objective variable). In unsupervised learning, the matrix decomposition method in the previous section is typical and it tries to extract features hidden in the data without using the given objective variable. Recently, reinforcement learning that maximizes rewards for behavior in a particular environment has been classified separately.

To date, several ML systems have been applied to brain image data of patients with neuropsychiatric disorders (Arbabshirani et al., 2017). Alzheimer's disease (AD) is clinically characterized by impaired memory and cognitive functions and it affects not only individuals but also society in general. AD is a neurodegenerative disease and its diagnosis can be improved by the use of biomarkers, especially with sMRI, to identify nerve loss; thus, it plays an important role in the clinical evaluation of patients with suspected AD. Through this image analysis, it has been shown that atrophy of local areas such as the hippocampus and the entorhinal cortex reflects the progress of the disease (Frisoni et al., 2010). Symptomatic treatment of AD is considered effective only when diagnosis and treatment return to the early progenitor stage. Predicting progression from AD to mild cognitive impairment (MCI), which is the intermediate condition between AD and normal, is useful for early detection and leads to effective treatment.

The most commonly used discrimination method is the support vector machine (SVM, Vapnik, 1999). However, since each method includes more dimensions that was originally assumed, it is difficult to directly apply the methods. This has been solved by devising appropriate preprocessing methods (Cuingnet et al., 2011). Vemuri et al. (2008) reduced the resolution to a voxel size of 8 mm on each side, and applied SVM only to $22 \times 27 \times 22$ voxels. In Fan et al. (2007), a method called COMPARE was developed, and SVM was used in conjunction with the standardization method HAMMER (Shen and Davatzikos, 2002). Otherwise, a region of interest (ROI) consisting of multiple voxels without inputting voxel values (Lerch et al., 2008, Magnin et al., 2009) or a principal component analysis wherein the main component is used (Teipel et al., 2007) can be considered as alternative methods. As these methods aggregate images, information may be lost.

On the other hand, Klöppel et al. (2008) proposed a discrimination method using gray matter by combining the high dimensional image standardization method DARTEL (Ashburner, 2007) and linear SVM to directly handle voxels. In addition, there are alternative methods to SVM, and regularized logistic regression models Casanova et al. (2011) are used in combination with the high dimensional image standardization method SyN (Avants et al., 2008), which can be executed with ANTS software. The PLS method was used for Phan et al. (2010). Basis expansion method is also effective as a dimension reduction method to reduce the number of input variables. Reiss and Ogden (2010b) uses radial basis functions on space, and Hackmack et al. (2012) considers a method using wavelets to support various resolutions.

The discrimination method in sMRI has applications to fields other than dementia research; for example, Ecker et al. (2010) applied SVM in the autism study. Yotter et al. (2011) combined fractal dimension and SVM to study multiple sclerosis. Weygandt et al. (2012) applied SVM and its ensemble learning to investigations into bulimia nervosa based on ROI volume.

In fMRI data analysis, multivoxel pattern analysis is used to summarize multivariate analytical methods, such as machine learning, using all voxels (O'Toole et al., 2007). Yamashita et al. (2008) and Ryali et al. (2010) applied a regularized logistic regression model. Casanova et al. (2011) used Random Forest and L1 regularized logistic regression analysis to examine the relationship between hormonal therapy and brain atrophy. Weygandt et al. (2012) used SVM and the searchlight method (Haynes et al., 2007; Kriegeskorte et al., 2006) to search for activation sites corresponding to obsessive compulsive disorder.

Support vector machine

The discrimination boundary of the logistic discrimination introduced earlier is a straight line. There are also quadratic discriminants in which the discriminant boundary is a quadratic curve. If you draw a more complicated discriminant boundary, you may be able to discriminate more accurately. Support vector machine (SVM) is one of the non-linear discriminants and has been studied and applied as a powerful discriminating method. As features, only observation values that are effective for discrimination (support vectors) are used. We introduce the concept of maximizing margins. The margin is the "distance between the discrimination boundary and the data." Also, consider projection to higher dimensions. It can be imagined that the discrimination ability is higher when viewing from the sky (three-dimensional) than when viewing on flat ground (two-dimensional). At that time, the discriminant boundary can be estimated by using the dual problem and the kernel trick as optimization problems. To give flexibility to the discriminant boundary, a soft margin (which may be a little inside the opponent) is introduced and is controlled by a parameter called a slack variable.

In order to distinguish group A and group B with $y = 1$ representing group A and $y = -1$ representing group B, the discrimination boundary is given as $f(x) = 0$ where

$$f(x) = w_0 + w_1 x_1 + \cdots + w_p x_p = w^\top x$$

and w is the weight vector which is estimated from the data. Discrimination method from the obtained $f(x)$

$$f(x) \geq 0 \Longrightarrow \text{Belongs to group A}$$
$$f(x) < 0 \Longrightarrow \text{Belongs to group B}$$

w is obtained from data (x_i, y_i) $(i = 1, 2, ..., n)$ so that the margin M is maximized. In general, the distance between a point and a boundary is given by

$$\frac{|f(x)|}{\|w\|} = \frac{|w^\top x|}{\|w\|}.$$

If the data of both groups are cleanly separated at the boundary (linear separability), then the margin M is defined as the distance from the boundary to the nearest data and for any i-th subject

$$\frac{w^\top x_i}{\|w\|} \geq M \quad (\text{if } y_i = 1), \qquad -\frac{w^\top x_i}{\|w\|} \geq M \quad (\text{if } y_i = -1)$$

and these can be summarized as follows.

$$\frac{y_i w^\top x_i}{\|w\|} \geq M \Rightarrow \frac{y_i w^\top x_i}{M \|w\|} \geq 1.$$

Replacing $w/M\|w\|$ with w, we obtain,

$$y_i w^\top x_i \geq 1. \tag{4.11}$$

If x_i is a support vector, then the equality holds, that is, $y_i w^\top x_i = 1$ and

$$M = \frac{y_i w^\top x_i}{\|w\|} = \frac{1}{\|w\|}. \tag{4.12}$$

w is optimized to maximize the margin in equation (4.12) under the conditions of (4.11), that is, the optimization problem is given as follows.

$$\max_{w} \frac{1}{\|w\|} \text{ subject to } y_i w^\top x_i \geq 1.$$

We relax the conditions of the linear separability (4.11) as follows.

$$y_i w^\top x_i \geq 1 - \xi_i,$$

for $\xi_i \geq 0$ $(i = 1, 2, ..., n)$. This ξ_i is called a slack variable and it allows for the crossing of boundaries as shown in Figure 4.6.

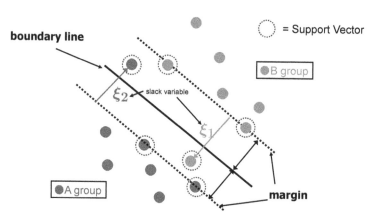

Figure 4.6: Discrimination boundary, margin and slack variable for SVM.

Since the solution cannot be found for any value of ξ_i, we put a constraint on margin maximization and the optimization problem is transformed as follows.

$$\max_{w} \frac{1}{\|w\|} + C \sum_{i=1}^{n} \xi_i \text{ subject to } y_i w^\top x_i \geq 1 - \xi_i \text{ and } \xi_i \geq 0$$

The constant C determines the balance between margin maximization and relaxation from full linear separability.

The higher the dimension, the better the classification accuracy. For example, it is easier to distinguish forest trees from above than from the plains. This can be done by transforming $f(x)$ as follows.

$$f(x) = w^\top x \Longrightarrow f(x) = v^\top \Phi(x)$$

where $\Phi(x)$ is a function that converts the dimensions of the input data to higher dimensions. As an example, the following function performs a two-dimensional to three-dimensional transformation.

$$\Phi(x) = \Phi(x_1, x_2) = (x_1^2, x_2^2, \sqrt{2} x_1 x_2)^\top.$$

How many dimensions do you project? Higher the dimension, higher is the likelihood of the discrimination system being true, hence the infinite dimension will be the best. However, can it be done? The kernel trick is a method to make the calculation easy while considering the projection to the infinite dimension and at the same time, also non-linearise the boundary line.

The final discriminant function in SVM,

$$f(\boldsymbol{x}) = \sum_{i=1}^{\#SV} \alpha_i y_i K(\boldsymbol{x}, \boldsymbol{x}_i) + b$$

where $K(\boldsymbol{a}, \boldsymbol{b})$ is a kernel function (of many types). The most representative is the Gaussian radial basis function (RBF) kernel,

$$K(\boldsymbol{a}, \boldsymbol{b}) = \exp(-\sigma \|\boldsymbol{a} - \boldsymbol{b}\|).$$

We also have polynomial, linear, sigmoid, Laplacian (RBF), Bessel, ANOVA RBF, spline and string kernels, respectively.

SVM has tuning parameters to smooth the discriminant boundary. These are the influence degree C of the slack variable and the σ in a Gaussian RBF kernel. These are selected by cross-validation or the information criterion. Various slack variables corresponding to $C = 1,10,100$ are illustrated in Figure 4.7 for tuning parameters and boundaries for SVM.

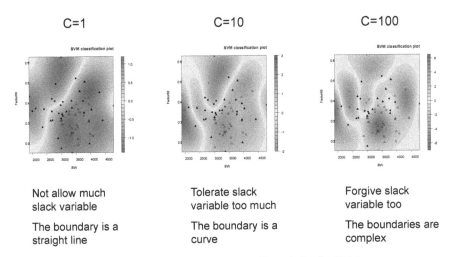

C=1	C=10	C=100
Not allow much slack variable	Tolerate slack variable too much	Forgive slack variable too
The boundary is a straight line	The boundary is a curve	The boundaries are complex

Figure 4.7: Tuning parameters and boundaries for SVM.

SVM is assertive whether it is group A or group B and cannot output the probability of belonging to group A like logistic discrimination. Therefore, Platt et al. (1999) considered using the following equation as the classification probability.

$$Pr(Y = 1 | f(\boldsymbol{x})) = \frac{1}{1 + \exp\{A\hat{f}(\boldsymbol{x}) + B\}}$$

where

$$\hat{f}(\boldsymbol{x}) = \sum_{i=1}^{\#SV} \hat{\alpha}_i y_i K(\boldsymbol{x}, \boldsymbol{x}_i) + \hat{b}.$$

A and B are determined to maximize the log-likelihood function from the logistic regression model with $\hat{f}(\boldsymbol{x})$ as an explanatory variable.

Tree model

The tree models perform discrimination or regression. The combination of explanatory variables related to the objective variable is represented by a tree diagram. The predicted values of the objective variable are displayed on the leaves of the tree diagram. More complex (interaction, non-linear) regression or discrimination can be performed, variable selection can be performed and cut-off values can be obtained for continuous variables. The tree diagram makes the results easier to interpret. In the algorithm of Tree model, for a parent node to which multiple subjects belong, we explore the branching conditions using variables and their cutoffs and divide the subjects into two child nodes. This branching continues from the root node to which all subjects belong, to the terminal node to which a predetermined minimum number of subjects belong. The following is the algorithm.

Algorithm

1. At the parent node, perform a "test" on all possible branches (all explanatory variables, possible values for each explanatory variable).

2. Branch at the "best" branch as a result of the "test" in 1. (Create child node)

3. Next, the child node obtained in step 2 is regarded as the parent node.

4. 1. 2. and 3. are performed until "branching is impossible."

The branch index is used to find the best branch that is also in the algorithm. At each node i, the following values (deviation, impurity) are calculated (K group discrimination problem).

■ Deviance

$$D_i = -2 \sum_{k=1}^{K} n_{ik} \log(p_{ik})$$

■ Entropy

$$D_i = - \sum_{k=1}^{K} p_{ik} \log(p_{ik})$$

■ Gini index

$$D_i = 1 - \sum_{k=1}^{K} p_{ik}^2$$

For the optimal branch, the degree of deviance and the change in impurity are calculated ("test") and the branch with the largest value is optimized.

$$Change \ = \ D(parent) - (D(\text{child left}) + D(\text{child right}))$$

$$Change \ = \ \text{likelihood ratio}\left(= \frac{\text{likelihood of parent}}{\text{likelihood of children}} \right)$$

In other words, when we consider a contingency table, we will choose a branch that makes the difference between the groups the most significant.

The non-branch in the algorithm determines how far the tree should grow, that is when to end the branching. This allows the analyst to give a condition for terminating the branch. For example, there are at least five nodes at the node to be branched, only a small number (0) at the terminal node and branching to a predetermined number of divisions (depth). These can be set by each software package.

Once a tree is completed, it may be pruned. Overgrown trees have many branches and are difficult to interpret, so the branches are reduced. This is called pruning. Pruning reduces "complexity." The resulting tree is denoted by T and its complexity is given by,

$$R_\alpha(T) = R(T) + \alpha \times |T|$$

where $R(T)$ is the cost (error, misclassification rate) and α is the complexity parameter (> 0) and $|T|$ is the number of terminal nodes. The tree keeps growing when $\alpha = 0$. As a result, the training is overfitted and is not suitable for prediction. α is determined to minimize the K-fold cross-validation (CV) error.

Random forest

If there are many trees, it becomes a forest. As shown in Figure 4.8, the random forest method is created with multiple tree models as described above and performs discrimination based on the set. To create multiple trees, the bootstrap method of the following procedure is used. Suppose there are n data (id = 1, 2,..., n). This removes n random integer values from 1 to n, allowing duplication. The individual value with the corresponding id is used as training data. Individual values not selected in this bootstrap are treated as Out of Bag (OOB) test data and used for performance evaluation. Trees in a random forest do not require pruning. Instead of all variables, m randomly selected variables are used as candidate variables for each branch. By default, $m =$ square root of the number of all variables used. The results of individual predictions from all the trees are decided by a majority in the case of discrimination, and by an average in the case of regression.

The characteristics of the random forest are as follows. (1) It is capable of capturing nonlinear structures (Tree also). (2) Since variables are selected randomly

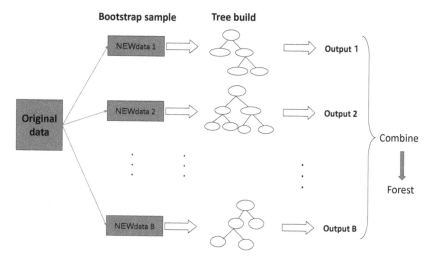

Figure 4.8: Image for the random forest method.

at the time of branching, they can be applied to hundreds or thousands of high-dimensional data. (3) The reliability of the RF model can be evaluated by estimating the OOB error.

The output of the random forest is the predicted value (similar to the tree). In the case of discrimination, it is the discriminated group and in the case of regression, it is the average value at that node. Additionally, there is a value called variable importance (VI) that indicates the importance of a variable in discrimination (regression). A higher value indicates that the variable plays a more important role in discrimination (regression) than other variables. Proximity is an index that indicates the closeness between individuals. It is used for substitution of missing values and detection of outliers.

The method of calculating VI is as follows. When the b-th tree is created, substitute the OOB sample into this tree and calculate the prediction error (misclassification rate). At this time, the values of the j-th variable are randomly rearranged in the OOB sample and the prediction error is calculated again. Compute the difference between the original prediction error and the re-arranged prediction error. The difference between all the B trees is averaged and defined as VI of the variable j. Therefore, VI measures the effect on the prediction error when the effect of the j-th variable is eliminated (random reordering). If you do the same for a linear regression model, the difference in prediction error will be the estimated regression coefficient ($\times 2$).

The calculation method of Proximity (similarity of the subject) is as follows. When there are n data, an $n \times n$ proximity matrix (symmetric matrix) is created from the random forest. If two sets of OOB samples i and j belong to the same terminal node in a tree in a random forest, 1 is counted in the i-th row and j-th column or the j-th row and i-th column of the proximity matrix. This is added to the B trees to create a proximity matrix.

Tuning parameters exist in the random forest as well as in the tuning parameter SVM (default value in randomForest of R). (1) Number of trees (500), (2) Number of branch candidate variables (square root of the number of variables), (3) Minimum number of people at terminal node.

R example

The supervised sparse principal component analysis is implemented as in the previous section and the score is extracted. The objective binary variable Z is transformed to the factor variable when Z=1 and control variable when Z=0 and stored as a dataset with the score.

```
fit113 = msma(SB1, Z=Z, comp=2, lambdaX=0.075, muX=0.5)
Ss = fit113$ssX
colnames(Ss) = paste("c", c(1:ncol(Ss)), sep="")
swdata113 = data.frame(
Z = as.factor(ifelse(Z == 1, "Y", "N")), Ss)
```

The resulting dataset is used to create a predictive model.

Logistic regression model

Fit the logistic model with the glm function and displaying the glm fit results.

```
glmfit = glm(Z~., data=swdata113, family=binomial)
summary(glmfit)
```

```
##
## Call:
## glm(formula = Z ~ ., family = binomial, data = swdata113)
##
## Deviance Residuals:
##          Min            1Q       Median            3Q
## -1.595e-04   -2.100e-08    0.000e+00    2.100e-08
##          Max
##    1.693e-04
##
## Coefficients:
```

```
##              Estimate Std. Error z value Pr(>|z|)
## (Intercept)    -7.743  30593.997   0.000    1.000
## c1            157.944  34585.053   0.005    0.996
## c2            -41.315  12290.486  -0.003    0.997
##
## (Dispersion parameter for binomial family taken to be 1)
##
##      Null deviance: 5.5452e+01  on 39  degrees of freedom
## Residual deviance: 5.4845e-08  on 37  degrees of freedom
## AIC: 6
##
## Number of Fisher Scoring iterations: 25
```

In this example, neither of the two components is significant by 5%, but a smaller p-value can be obtained if only C1 is used. Convergence does not seem to have worked (the number of iterations: 25).

Next the diagnostic probabilities are computed and are transformed into a binary variable that may or may not be greater than or equal to 0.5.

```
test = predict(glmfit, type="response")>=0.5
```

Creation of a confusion table (input and output are performed simultaneously by enclosing in parentheses).

```
(err.table = table(swdata113$Z, test))
```

```
##     test
##      FALSE TRUE
##   N    20    0
##   Y     0   20
```

The confusion matrix can be calculated as follows to obtain the classification error rate, that is, it is the proportion of the non-diagonal component.

```
1 - sum(diag(err.table)) / sum(err.table)
```

```
## [1] 0
```

The discrimination is perfect. The significance of the coefficients and discriminant performance seem to be irrelevant. However, this is a result of training data alone, and discriminant ability should be measured by independent test data. This will be discussed in a later section.

The confusion matrix can also be calculated as follows to obtain sensitivity, specificity, false positive rate and false negative rate.

```
t(apply(err.table, 1, function(x) x / sum(x)))
```

```
##    test
##     FALSE TRUE
## N    1    0
## Y    0    1
```

In this example, it is arranged as follows.

```
matrix(
c("specificity", "false positive rate",
"false negative rate","sensitivity")
, ncol=2)
```

```
##       [,1]                 [,2]
## [1,] "specificity"        "false negative rate"
## [2,] "false positive rate" "sensitivity"
```

Next, we draw a probability plot with component 1 score on the horizontal axis and component 2 score on the vertical axis, and plot the probability values in different colors.

```
x = seq(min(swdata113$c1), max(swdata113$c1), length = 30)
y = seq(min(swdata113$c2), max(swdata113$c2),
length = length(x))
prob = function(x, y) 1/(1+exp(-predict(glmfit,
newdata=data.frame(c1=x, c2=y))))
z = outer(x, y, prob)
filled.contour(x,y,z, xlab="Component 1", ylab="Component 2")
```

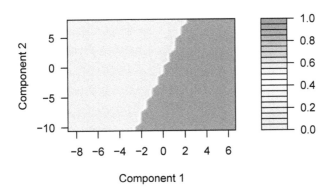

Since the boundary line with a probability of 0.5 is almost perpendicular to the abscissa, it seems that component 1 is more effective for discrimination.

Support vector machine

To run the support vector machine, load the `e1071` package.

```
library(e1071)
```

The `tune` function of the `e1071` package is used to find the optimal tuning parameters. The tuning parameters are σ in the Gaussian kernel (`gamma`) and the influence degree C of the slack variable (`cost`). By setting the cross to the sample size, the leave-one-out cross-validation method is used, and the classification error is used as a performance indicator.

```
set.seed(1)
tuneSVM = tune(svm, Z~., data=swdata113,
ranges = list(gamma = 2^(0:2), cost = c(4, 6, 8)),
tunecontrol = tune.control(cross = nrow(swdata113)))
```

The results of the tuning (table and plot) are as follows.

```
summary(tuneSVM)
```

```
##
## Parameter tuning of 'svm':
##
## - sampling method: leave-one-out
##
## - best parameters:
##   gamma cost
##      2    4
##
## - best performance: 0.05
##
## - Detailed performance results:
##    gamma cost error dispersion
## 1      1    4 0.075  0.2667468
## 2      2    4 0.050  0.2207214
## 3      4    4 0.050  0.2207214
## 4      1    6 0.075  0.2667468
## 5      2    6 0.050  0.2207214
## 6      4    6 0.050  0.2207214
## 7      1    8 0.050  0.2207214
## 8      2    8 0.050  0.2207214
## 9      4    8 0.075  0.2667468
```

A grid search was performed to calculate the errors for all combinations of tuning parameters and to optimize the parameters with their minimum values. With tuning parameters on the horizontal and vertical axes, a color-coded plot of the classification errors is drawn as follows.

```
plot(tuneSVM, color.palette = heat.colors)
```

Performance of `svm'

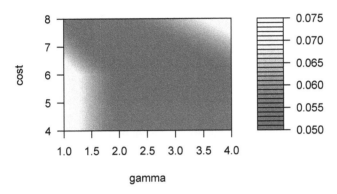

gamma

The optimal parameters can be extracted as follows.

```
bestGamma = tuneSVM$best.parameters$gamma
bestC = tuneSVM$best.parameters$cost
```

The fit of the SVM with the optimal parameters is performed.

```
set.seed(1)
svmfit = svm(Z~., data=swdata113,
cost = bestC, gamma = bestGamma,
probability=TRUE, kernel="radial", cross=nrow(swdata113))
summary(svmfit)
```

```
##
## Call:
## svm(formula = Z ~ ., data = swdata113, cost = bestC,
##      gamma = bestGamma, probability = TRUE, kernel = "radial",
##      cross = nrow(swdata113))
##
##
## Parameters:
##    SVM-Type:  C-classification
## SVM-Kernel:  radial
##        cost:  4
##
## Number of Support Vectors:  22
##
## ( 11 11 )
##
```

```
##
## Number of Classes:  2
##
## Levels:
##  N Y
##
## 40-fold cross-validation on training data:
##
## Total Accuracy: 95
## Single Accuracies:
##   100 100 100 100 100 0 100 0 100 100 100 100 100 100 100
      100 100 100 100 100 100 100 100 100 100 100 100 100 100
      100 100 100 100 100 100 100 100 100 100 100
```

Similar to logistic classification, the discrimination results by the SVM, confusion matrices and the classification errors are computed.

```
pred = predict(svmfit, newdata=swdata113, probability=TRUE,
decision.values=TRUE)
(err.table = table(swdata113$Z, pred))
```

```
##     pred
##      N  Y
##   N 20  0
##   Y  0 20
```

```
1 - sum(diag(err.table)) / sum(err.table)
```

```
## [1] 0
```

The plot of the discriminant boundary is provided in the e1071 package with a special function.

```
plot(svmfit, swdata113, c2~c1)
```

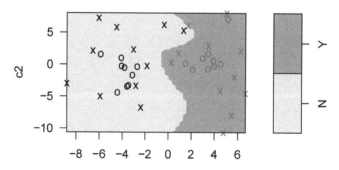

SVM classification plot

It shows that the discrimination boundaries are nonlinear. Furthermore, as in the case of logistic classification, the component 1 score is valid for discrimination because it is almost perpendicular to the vertical axis.

Tree model

The tree model is used for discrimination. Since the tree model is more effective with more variables, the number of components is increased to 20.

```
opt11 = optparasearch(SB1, Z=Z, comp=20,
search.method = "regparaonly", criterion="BIC")
(fit311 = msma(SB1, Z=Z,
comp=opt11$optncomp, lambdaX=opt11$optlambdaX))
```

```
## Call:
## msma.default(X = SB1, Z = Z, comp = opt11$optncomp,
    lambdaX = opt11$optlambdaX)
##
## Numbers of non-zeros for X:
##          comp1 comp2 comp3 comp4 comp5 comp6 comp7 comp8
## block1      16    38    31    20    24    33    19    20
##          comp9 comp10 comp11 comp12 comp13 comp14 comp15
## block1      23     26     32     24     22     32     30
##          comp16 comp17 comp18 comp19 comp20
## block1      22     20     18     29     35
##
## Numbers of non-zeros for X super:
##   comp1  comp2  comp3  comp4  comp5  comp6  comp7
##       1      1      1      1      1      1      1
##   comp8  comp9 comp10 comp11 comp12 comp13 comp14
##       1      1      1      1      1      1      1
## comp15 comp16 comp17 comp18 comp19 comp20
##       1      1      1      1      1      1
```

```
Ss = fit311$ssX
colnames(Ss) = paste("c", c(1:ncol(Ss)), sep="")
swdata311 = data.frame(
Z = as.factor(ifelse(Z == 1, "Y", "N")), Ss)
```

The rpart package to run the tree model and the rpart.plot package to illustrate it are loaded.

```
library(rpart)
library(rpart.plot)
```

Given the seed of the random number (to fix the result of CV), the tree model is fitted (default is Gini index) by using the `rpart` function. The minimum number of cases required in a node to be divided was set to 4 (= `minsplit`).

```
set.seed(1)
(treefit = rpart(Z~., data=swdata311,
control = rpart.control(minsplit = 4)))
```

```
## n= 40
##
## node), split, n, loss, yval, (yprob)
##       * denotes terminal node
##
##   1) root 40 20 N (0.50000000 0.50000000)
##     2) c6< 0.6568338 19  1 N (0.94736842 0.05263158)
##       4) c1>=-5.036697 18  0 N (1.00000000 0.00000000) *
##       5) c1< -5.036697 1   0 Y (0.00000000 1.00000000) *
##     3) c6>=0.6568338 21  2 Y (0.09523810 0.90476190)
##       6) c1>=4.308694 1   0 N (1.00000000 0.00000000) *
##       7) c1< 4.308694 20  1 Y (0.05000000 0.95000000)
##         14) c14< -6.065711 1   0 N (1.00000000 0.00000000) *
##         15) c14>=-6.065711 19  0 Y (0.00000000 1.00000000) *
```

The tree is plotted. It is possible to use the function `prp` to draw a better looking tree than the default plot function, such as a constant width between nodal points, number of individuals per group at branch points and all branch points.

```
prp(treefit, type=4, extra=1, faclen=0, nn=TRUE)
```

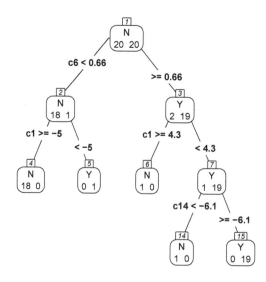

The rpart function records CV classification errors for each complexity parameter corresponding to the number of splits (nsplit) when it is executed. It can
be shown as follows.

```
printcp(treefit)
```

```
##
## Classification tree:
## rpart(formula = Z ~ ., data = swdata311,
    control = rpart.control(minsplit = 4))
##
## Variables actually used in tree construction:
## [1] c1   c14 c6
##
## Root node error: 20/40 = 0.5
##
## n= 40
##
##      CP nsplit rel error xerror    xstd
## 1 0.85      0     1.00    1.30 0.15083
## 2 0.05      1     0.15    0.25 0.10458
## 3 0.01      4     0.00    0.30 0.11292
```

This result is plotted as follows, where size of tree is expressed in terms of the
number of leaves.

```
plotcp(treefit)
```

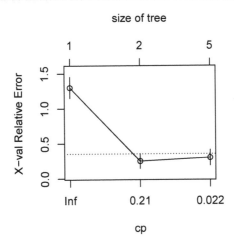

These results indicate that the number of leaves is 2, i.e., up to the first bifurca-
tion, for good classification.

Pruning to reduce wasteful branching is performed by setting the cp value as follows.

```
(treefit1 = prune(treefit, cp=0.05))
```

```
## n= 40
##
## node), split, n, loss, yval, (yprob)
##       * denotes terminal node
##
## 1) root 40 20 N (0.50000000 0.50000000)
##    2) c6< 0.6568338 19  1 N (0.94736842 0.05263158) *
##    3) c6>=0.6568338 21  2 Y (0.09523810 0.90476190) *
```

It is also possible to specify a leaf number and prune below it.

```
(treefit2 = snip.rpart(treefit, 7))
```

```
## n= 40
##
## node), split, n, loss, yval, (yprob)
##       * denotes terminal node
##
## 1) root 40 20 N (0.50000000 0.50000000)
##    2) c6< 0.6568338 19  1 N (0.94736842 0.05263158)
##      4) c1>=-5.036697 18  0 N (1.00000000 0.00000000) *
##      5) c1< -5.036697 1  0 Y (0.00000000 1.00000000) *
##    3) c6>=0.6568338 21  2 Y (0.09523810 0.90476190)
##      6) c1>=4.308694 1  0 N (1.00000000 0.00000000) *
##      7) c1< 4.308694 20  1 Y (0.05000000 0.95000000) *
```

Graphic representation of a pruned Tree.

```
prp(treefit2, type=4, extra=1, faclen=0, nn=TRUE)
```

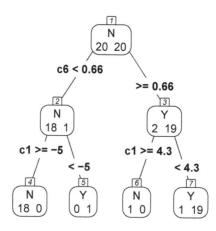

Predict training data. Specify type and set for discrimination.

```
pred = predict(treefit2, type="class")
```

Create a table of discrimination results and labels.

```
(err.table = table(swdata311$Z, pred))
```

```
##      pred
##       N  Y
##   N  19  1
##   Y   0 20
```

Calculation of classification error rate.

```
1 - sum(diag(err.table))/sum(err.table)
```

```
## [1] 0.025
```

Only one case was misclassified.

Random forests

To begin, we prepare a partial dataset of only six cases to illustrate the bootstrapping method used in random forests.

```
swdata3112 = head(swdata311)
```

Set the seed of the random number, id=1,2,...,6 to extract six numbers at random, allowing for duplication.

```
set.seed(1)
(idrand = sample(1:6, replace=TRUE))
```

```
## [1] 2 3 4 6 2 6
```

The rows corresponding to these numbers are taken from the dataset, and the resulting is the bootstrap dataset.

```
(bsample = swdata3112[idrand, 1:4])
```

```
##     Z         c1         c2        c3
## 2   N  3.0211928  0.6256017  3.110181
## 3   N  0.6568067  2.1047445  5.726669
## 4   N  2.0028711 -4.7660248 -2.746751
## 6   N -4.6606108 -1.8953655 -5.161246
## 2.1 N  3.0211928  0.6256017  3.110181
## 6.1 N -4.6606108 -1.8953655 -5.161246
```

The OOB dataset is composed of rows not included in the bootstrap dataset.

```
(oobsample = swdata3112[!(1:nrow(swdata3112) %in%
unique(idrand)), 1:4])
```

```
##   Z       c1         c2        c3
## 1 N 1.582432 -3.163756 -5.552536
## 5 N 3.236069  8.255753 -1.725270
```

Load randomForest package and the e1071 package for the tuning selection (assuming it is installed).

```
library(randomForest)
library(e1071)
```

Set random number seed, the selection of tuning parameters is implemented by leave-one-out CV using the tune function of the e1071 package.

```
set.seed(1)
tuneRF = tune(randomForest, Z~., data=swdata311,
ranges = list(mtry = c(4,6,8), ntree = c(300, 500, 1000),
nodesize= c(1,2,3)),
tunecontrol = tune.control(cross = nrow(swdata311)))
```

Tuning parameters are the number of candidate variables for the split (mtry), the number of trees (ntree) and the node size for each tree (nodesize).

The results of the tuning (table and plot) are as follows.

```
summary(tuneRF)
```

```
##
## Parameter tuning of 'randomForest':
##
## - sampling method: leave-one-out
##
## - best parameters:
##  mtry ntree nodesize
##     6   300        1
##
## - best performance: 0.125
##
## - Detailed performance results:
##    mtry ntree nodesize error dispersion
## 1     4   300        1 0.150  0.3616203
## 2     6   300        1 0.125  0.3349321
## 3     8   300        1 0.150  0.3616203
```

```
## 4     4    500      1 0.150   0.3616203
## 5     6    500      1 0.150   0.3616203
## 6     8    500      1 0.150   0.3616203
## 7     4   1000      1 0.150   0.3616203
## 8     6   1000      1 0.150   0.3616203
## 9     8   1000      1 0.150   0.3616203
## 10    4    300      2 0.125   0.3349321
## 11    6    300      2 0.150   0.3616203
## 12    8    300      2 0.150   0.3616203
## 13    4    500      2 0.150   0.3616203
## 14    6    500      2 0.150   0.3616203
## 15    8    500      2 0.150   0.3616203
## 16    4   1000      2 0.150   0.3616203
## 17    6   1000      2 0.150   0.3616203
## 18    8   1000      2 0.150   0.3616203
## 19    4    300      3 0.150   0.3616203
## 20    6    300      3 0.125   0.3349321
## 21    8    300      3 0.150   0.3616203
## 22    4    500      3 0.150   0.3616203
## 23    6    500      3 0.150   0.3616203
## 24    8    500      3 0.150   0.3616203
## 25    4   1000      3 0.150   0.3616203
## 26    6   1000      3 0.150   0.3616203
## 27    8   1000      3 0.150   0.3616203
```

The optimized tuning parameters are retrieved in the following manner.

```
bestmtry = tuneRF$best.parameters$mtry
bestntree = tuneRF$best.parameters$ntree
bestnodesize = tuneRF$best.parameters$nodesize
```

Setting a random number seed, and the random forest using selected parameters is executed. The proximity is also calculated.

```
set.seed(1)
(rffit = randomForest(Z~., data=swdata311, proximity=TRUE,
mtry = bestmtry, ntree = bestntree, nodesize=bestnodesize))

##
## Call:
##  randomForest(formula = Z ~ ., data = swdata311,
    proximity = TRUE, mtry = bestmtry, ntree = bestntree,
    nodesize = bestnodesize)
##              Type of random forest: classification
##                  Number of trees: 300
```

```
## No. of variables tried at each split: 6
##
##              OOB estimate of  error rate: 15%
## Confusion matrix:
##      N  Y class.error
## N 17  3         0.15
## Y  3 17         0.15
```

The Variable importance (VI)s are plotted in the order of their values.

```
varImpPlot(rffit)
```

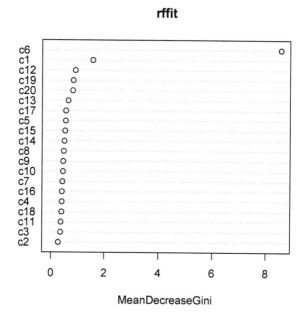

As in the Tree model, we found that component 6 was the most important in the model.

The loading for the component 6 was overlaid on the brain image.

```
Q = fit311$wbX[[1]][,6]
outstat1 = rec(Q, img1$imagedim, B=B1, mask=img1$brainpos)
outstat2 = -outstat1
coat(template, outstat2)
```

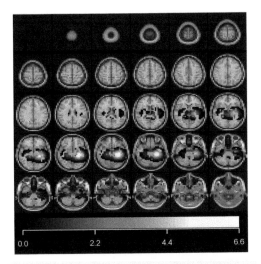

```
atlastable(tmpatlas, outstat2, atlasdataset)
```

```
##     ROIid                   ROIname sizepct sumvalue Min.
## 37     37        Left Hippocampus   1.000  174.526   0
## 39     39    Left Parahippocampus   1.000  146.409   0
## 55     55            Left Fusiform   0.976  150.882   0
## 89     89    Left Inferior Temporal 1.000  124.410   0
## 95     95      Left Cerebellum 4-5   1.000   65.705   0
## 77     77            Left Thalamus   1.000   59.129   0
## 47     47             Left Lingual   0.773   38.494   0
## 73     73             Left Putamen   1.000   26.886   0
## 93     93        Left Cerebellum 3   1.000   13.037   0
## 97     97        Left Cerebellum 6   1.000   34.237   0
##      Mean  Max.
## 37 0.005 6.601
## 39 0.005 6.532
## 55 0.005 5.864
## 89 0.004 4.502
## 95 0.002 4.163
## 77 0.002 3.917
## 47 0.001 3.815
## 73 0.001 3.690
## 93 0.000 3.661
## 97 0.001 3.163
```

This component represents the area around the left hippocampus.

The prediction on training data is implemented with specifying type of discrimination.

```
pred = predict(rffit, type="class")
```

A confusion matrix of discrimination results and labels is created.

```
(err.table = table(swdata311$Z, pred))
```

```
##      pred
##       N  Y
##   N  17  3
##   Y   3 17
```

Calculation of misclassification rate.

```
1 - sum(diag(err.table))/sum(err.table)
```

```
## [1] 0.15
```

In this case, six cases were misclassified.

Calculation of sensitivity, specificity, false positive, false negative

```
t(apply(err.table, 1, function(x) x / sum(x)))
```

```
##      pred
##         N     Y
##   N  0.85  0.15
##   Y  0.15  0.85
```

To illustrate how each explanatory variable is affected, a partial plot can be made as follows.

```
par(mfrow=c(1,2))
partialPlot(rffit, swdata311, c1)
partialPlot(rffit, swdata311, c2)
```

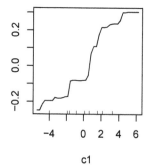
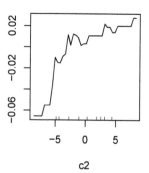

The proximity plot, which represents the similarity between the cases, is drawn as follows.

```
par(mfrow=c(1,1), mar=c(3,3,3,8))
z = rffit$proximity
n = nrow(swdata311)
filled.contour(x=1:n, y=1:n, z=z, color = terrain.colors)
```

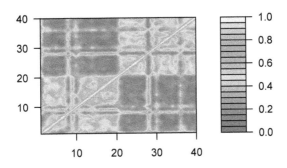

The result was that the first 10 cases were similar to each other and the next 10 cases were also similar to each other. The following two analyses can be performed from this proximity.

The multidimensional scaling (MDS) method, which is a method of arranging the similarities between individuals, with those that are similar in two-dimensional space close together and those that are far apart, is performed in the following way.

```
par(mfrow=c(1,1), mar=c(4,3,2,2))
MDSplot(rffit, factor(swdata311$Z),
pch=as.numeric(swdata311$Z)-1)
```

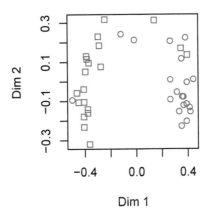

In this two-dimensional plot, individuals have different markers for each group. This figure also shows that the three members of each group are in positions where they are likely to be misclassified.

The outlier degree is calculated from the inverse of the mean value of each individual proximity and plotted as follows.

```
par(mfrow=c(1,1), mar=c(4,3,2,2))
plot(randomForest::outlier(rffit), type="h",
col= as.numeric(swdata311$Z))
```

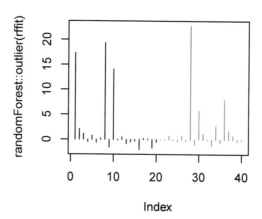

The results also show that there are three cases in each group that are different from the others.

Evaluation

Evaluation criteria

The confusion matrix is a basic matrix when considering the evaluation of the classification model and expresses the relationship between the predicted value of the model and the observed value. Specifically, there are four categories, true positive, true negative, false positive and false negative, as shown in Table 4.1.

Table 4.1: Confusion matrix for classification evaluation.

		Predicted	
		Positive	Negative
Actual	Positive	True Positive (TP)	False Negative (FN)
	Negative	False Positive (FP)	True Negative (TN)

The accuracy rate (accuracy) is calculated by the following formula and is the overall rate of successful prediction.

$$accuracy = \frac{TP + TN}{TP + FP + FN + TN}$$

The sensitivity (recall rate) is the proportion of the actual positive ones predicted to be positive. The specificity is the proportion of the actual negative ones predicted to be negative. The sensitivity and specificity are calculated by the following formula.

$$sensitivity(recall) = \frac{TP}{TP + FN}, \quad specificity = \frac{TN}{TN + FP}$$

The precision is calculated by the following formula and is the proportion of data that is positive among the data predicted to be positive.

$$precision = \frac{TP}{TP + FP}$$

The recall rate (recall) is calculated by the following formula and is the proportion of the actual positive ones predicted to be positive.

$$recall = \frac{TP}{TP + FN}$$

When evaluating the score of a continuous quantity, a receiver operator characteristic (ROC) curve is used. In the case of machine learning discrimination per-formance, the output is the prediction probability. The ROC curve gives a certain cutoff value to the continuous quantity, binaries, whether the machine-learning prediction is positive or negative and then calculates sensitivity and specificity. Multiple sensitivity and specificity can be obtained by changing the cutoff value. The ROC curve is plotted with 1-specificity (FPR) on the horizontal axis and sensitivity (TPR) on the vertical axis. The larger the area under the curve (AUC), the higher the performance.

These metrics can provide inaccurate and misleading information about classification performance for datasets with an imbalanced ratio of positive to negative subjects. For example, suppose there is test data of 1 case of positive and 99 cases of negative. If all 100 discrimination results are negative at this time, $Accuracy = \frac{99}{100} = 99\%$, which seems to be a good result. However, the positive score is 0%, which does not reflect the lack of positive detection. In the first place, the discrimination method that makes all negatives is a low-performance discrimination method that is possible even if it is not an advanced discrimination method and an index that correctly evaluates it is necessary.

In such a case, if recall or precision varies depending on the number of positives, this problem is overcome and it is more suitable for the evaluation of imbalance data. The PR (Precision-Recall) curve is a curve drawn by applying multiple cut-off values to the output of a continuous amount similar to the ROC curve, with Recall on the horizontal axis and Precision on the vertical axis. The AUC is calculated and evaluated by imbalanced data. Additionally, an F1 score (F1) defined as follows is also used as an evaluation index of imbalance.

$$F1 = \frac{2TP}{2TP + FP + FN}$$

Cross-validation

Even if the performance of the training data is improved by adjusting the tuning parameters or using a large number of features, the system may have fallen into "overfitting" which is effective only for some data. In the case of over-learning, the prediction performance of test data is unlikely to increase. "Validation" generally means "verification of model generalization performance." Generalization performance refers to "performance for unknown data (label)." Evaluation of generalization performance is important because the original purpose of the analysis is to discriminate unlabeled data. It is considered that a model with smaller fluctuations when the training data changes will have less dependence on the training data, thereby leading to better prediction. Therefore, evaluation with multiple training data is also necessary. Ideally, it is desirable to have multiple training data to make a model and then to evaluate it using completely independent test data. This is difficult when the number of subjects is limited. Furthermore, reliability is higher when all available data are used to determine the final model. In such a case, a method called cross-validation is used.

Cross-validation is a method of dividing training data and validation data for evaluation, creating test data from the data at hand and evaluating discrimination performance. It is a method of evaluating generalization performance that does not rely on specific data. Since the test data is based on the data at hand, cross-validation is also called internal validation and the method of preparing and evaluating completely independent-test data is called external validation. The evaluation is performed in the feature selection, tuning parameter selection and the performance of the final model. The cross-validation error is calculated as follows. In the data of *n* people at hand, *n* − 1 of them are treated as training data and one is treated as test data (the leave-one-out method). A discriminant function is created from the training data, the data of the remaining one person who became the test data is discriminated and the error (correct or incorrect) is recorded (creating a discrimination table). The above is repeated *n* times, and the cross-validation

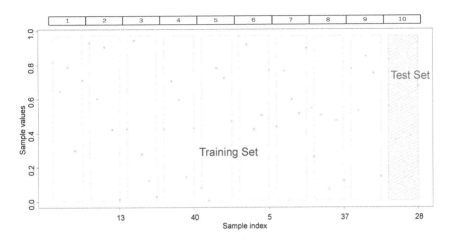

Figure 4.9: 10-fold cross-validation.

error is calculated by (number of errors)/n. Leave-one-out cross-validation is effective when n is small. However, when n is large, the number of verifications increases and the computation time becomes enormous. Furthermore, generalization may not be fully evaluated due to the high proportion of training data to test data. As a method for solving the problem, there is a technique called k-fold cross-validation. The method is described in the often occurring case of $k = 10$. As in Figure 4.9, the data set is randomly divided; 90% becoming the training data set and 10% the test data set. As leave-one-out cross-validation, the model is trained to use the training data set and evaluating the performance of the model using the test data set. Repeat this process 10 times to average the performance.

In k-fold cross-validation, the computational complexity is smaller than in leave-one-out and test data consisting of multiple subjects can be obtained, so it may be good for general-purpose evaluation, but there are some limitations. First, the data set is randomly divided into k pieces, only for the k pattern model and evaluating its versatility. A smaller value of "k" results in a smaller number of models and evaluations. Then the evaluation may be insufficient. Additionally, if k is small, the number of subjects in the training data is small and the reliability of the model is low. Conversely, in the leave-one-out cross-validation where $k = n$, there is no need for concern, but if k is large, the computational complexity increases and the amount of test data decreases as described above.

In consideration of these factors, it is common to select either $k = 10$ or $k = 5$ according to the total number of subjects n. Furthermore, repeated k-fold cross-validation that repeats randomly dividing the data set may be more effective. Since this is close to the OOB solution used in the random forest method, it can be concluded that the evaluation is based on the bootstrap method.

R example

Phenotypes such as disease status are identified by the regression model from brain image data. There are conventional functions in the Classification And REgression Training (`caret`) package that evaluate the predictive performance of this model.

For external verification, the test data with 500 subjects in one group independent of training data is generated by the caret function as follows.

```
img2 = simbrain(baseimg = baseimg, diffimg = diffimg2,
sdevimg=sdevimg, mask=mask, n0=500, c1=0.01, sd1=0.1,
zeromask=FALSE, seed=2)
```

The binary outcome should be converted into the factor to use the `caret` package.

```
testZ = as.factor(ifelse(img2$Z == 1, "Y", "N"))
```

The dimensions of the image data are reduced by the basis function.

```
SB2 = basisprod(img2$S, B1)
```

Logistic regression model

The `ptest` function is based on the caret package and uses the output of the `msma` function to fit the classification model described in the previous section. The logistic regression model is implemented with the argument regmethod = "glm" and the 5 repeated 10-fold cross validation is performed by default settings.

```
ptest113 = ptest(object=fit113, Z=Z, newdata=SB2,
testZ=testZ, regmethod = "glm")
```

The following are the fitting results of the logistic regression.

```
summary(ptest113$trainout$finalModel)
```

```
##
## Call:
## NULL
##
## Deviance Residuals:
##         Min          1Q       Median          3Q
## -1.595e-04  -2.100e-08   0.000e+00   2.100e-08
##         Max
##   1.693e-04
##
## Coefficients:
##                 Estimate Std. Error z value Pr(>|z|)
```

```
## (Intercept)    -7.743  30593.997   0.000   1.000
## c1            157.944  34585.053   0.005   0.996
## c2            -41.315  12290.486  -0.003   0.997
##
## (Dispersion parameter for binomial family taken to be 1)
##
##      Null deviance: 5.5452e+01  on 39  degrees of freedom
## Residual deviance: 5.4845e-08  on 37  degrees of freedom
## AIC: 6
##
## Number of Fisher Scoring iterations: 25
```

Here's a summary of the evaluation methods and training data results.

```
ptest113$trainout
```

```
## Generalized Linear Model
##
## 40 samples
##  2 predictor
##  2 classes: 'N', 'Y'
##
## No pre-processing
## Resampling: Cross-Validated (10 fold, repeated 5 times)
## Summary of sample sizes: 36, 36, 36, 36, 36, 36, ...
## Resampling results:
##
##   ROC     Sens  Spec  Accuracy  Kappa  AUC  Precision
##   0.9875  0.94  0.91  0.925     0.85   NaN  0.9366667
##   Recall  F
##   0.94    0.9246667
```

The model also evaluates the prediction performance of the test data (the dimension reduced SB2 and the testZ).

```
ptest113$predcnfmat
```

```
## Confusion Matrix and Statistics
##
##           Reference
## Prediction   N   Y
##          N 408 102
##          Y  92 398
##
##                 Accuracy : 0.806
```

```
##                        95% CI : (0.7801, 0.8301)
##      No Information Rate : 0.5
##      P-Value [Acc > NIR] : <2e-16
##
##                        Kappa : 0.612
##
##   Mcnemar's Test P-Value : 0.5182
##
##             Sensitivity : 0.7960
##             Specificity : 0.8160
##          Pos Pred Value : 0.8122
##          Neg Pred Value : 0.8000
##               Precision : 0.8122
##                  Recall : 0.7960
##                      F1 : 0.8040
##              Prevalence : 0.5000
##          Detection Rate : 0.3980
##    Detection Prevalence : 0.4900
##       Balanced Accuracy : 0.8060
##
##         'Positive' Class : Y
##
```

In comparison, the output of msma function with $\mu = 0$ is applied to the ptest function.

```
ptest112 = ptest(object=fit112, Z=Z, newdata=SB2, testZ=testZ,
regmethod = "glm")
ptest112$predcnfmat$overall["Accuracy"]
```

```
## Accuracy
##     0.524
```

The prediction accuracy was 0.806 for the method with the supervised PCA and 0.524 for the original (unsupervised) PCA. The supervised PCA overperformed.

The input of the ptest function can be a specified data matrix instead of the msma fit.

```
ptest311 = ptest(object=img1$S, Z=Z, newdata=img2$S,
testZ=testZ, regmethod = "glm")
ptest312 = ptest(object=SB1, Z=Z, newdata=SB2, testZ=testZ,
regmethod = "glm")
```

The accuracy of the prediction based on the original data was slightly worse than that of the dimension reduced method.

```
ptest311$predcnfmat$overall["Accuracy"]
```

```
## Accuracy
##    0.495
```

```
ptest312$predcnfmat$overall["Accuracy"]
```

```
## Accuracy
##    0.478
```

From these results, it was found that the prediction accuracy was not high in the original image data matrix or by applying only basis functions.

The supervision amount is controlled by the parameter $0 \leq \mu \leq 1$. The larger value indicates stronger supervision (strongly correlated with outcome Z). The effect of parameter μ is examined below.

```
mus = seq(0, 1, by=0.25)
comps = c(1,2,5,10)
```

```
paramtest1 = lapply(comps, function(c1){lapply(mus,
function(mu1){
tmpfit = msma(SB1, Z=Z, comp=c1, lambdaX=0.075, muX=mu1)
tmpptest = ptest(object=tmpfit, Z=Z, newdata=SB2,
testZ=testZ, regmethod = "glm")
tmpptest$predcnfmat
})})
```

```
out1 = do.call(rbind, lapply(paramtest1, function(x)
do.call(cbind, lapply(x,
function(y)y$overall["Accuracy"]))))
rownames(out1)=comps; colnames(out1)=mus
```

The result shows the number of components in the row and the parameter μ values in the column.

```
kable(out1, "latex", booktabs = T)
```

	0	0.25	0.5	0.75	1
1	0.532	0.823	0.825	0.820	0.830
2	0.524	0.818	0.806	0.812	0.815
5	0.766	0.786	0.767	0.804	0.775
10	0.712	0.755	0.772	0.789	0.772

SVM

SVM is evaluated with regmethod = "svmRadial". The candidate tuning parameters are specified in the same way as in the previous section.

```
svmgrid = expand.grid(sigma = c(0.001, 0.025, 0.05),
C = c(0.5, 0.75, 1))
ptest211 = ptest(object=fit311, Z=Z, newdata=SB2, testZ=testZ,
regmethod = "svmRadial", metric="ROC",
param=svmgrid)
ptest211$trainout
```

```
## Support Vector Machines with Radial Basis Function Kernel
##
## 40 samples
## 20 predictors
##  2 classes: 'N', 'Y'
##
## No pre-processing
## Resampling: Cross-Validated (10 fold, repeated 5 times)
## Summary of sample sizes: 36, 36, 36, 36, 36, 36, ...
## Resampling results across tuning parameters:
##
```

	sigma	C	ROC	Sens	Spec	Accuracy	Kappa	AUC
##	0.001	0.50	0.940	0.70	0.89	0.795	0.59	NaN
##	0.001	0.75	0.940	0.76	0.86	0.810	0.62	NaN
##	0.001	1.00	0.940	0.73	0.85	0.790	0.58	NaN
##	0.025	0.50	0.925	0.81	0.83	0.820	0.64	NaN
##	0.025	0.75	0.925	0.82	0.81	0.815	0.63	NaN
##	0.025	1.00	0.910	0.74	0.88	0.810	0.62	NaN
##	0.050	0.50	0.900	0.79	0.76	0.775	0.55	NaN
##	0.050	0.75	0.900	0.72	0.80	0.760	0.52	NaN
##	0.050	1.00	0.910	0.69	0.83	0.760	0.52	NaN

	Precision	Recall	F
##	0.9149660	0.70	0.7666667
##	0.8945578	0.76	0.7972789
##	0.8804348	0.73	0.7992754
##	0.8733333	0.81	0.8120000
##	0.8733333	0.82	0.8106667
##	0.9042553	0.74	0.8092199
##	0.8333333	0.79	0.7720000
##	0.8368056	0.72	0.7702128
##	0.8510638	0.69	0.7695652

```
##
## ROC was used to select the optimal model using
##  the largest value.
## The final values used for the model were sigma =
##  0.001 and C = 0.5.
```

After the tuning parameters have been selected, the confusion matrix and predictive evaluation index for the test data are computed.

```
ptest211$predcnfmat
```

```
## Confusion Matrix and Statistics
##
##               Reference
## Prediction    N    Y
##          N 216  411
##          Y 284   89
##
##                 Accuracy : 0.305
##                   95% CI : (0.2766, 0.3346)
##      No Information Rate : 0.5
##      P-Value [Acc > NIR] : 1
##
##                    Kappa : -0.39
##
##   Mcnemar's Test P-Value : 1.758e-06
##
##              Sensitivity : 0.1780
##              Specificity : 0.4320
##           Pos Pred Value : 0.2386
##           Neg Pred Value : 0.3445
##                Precision : 0.2386
##                   Recall : 0.1780
##                       F1 : 0.2039
##               Prevalence : 0.5000
##           Detection Rate : 0.0890
##     Detection Prevalence : 0.3730
##        Balanced Accuracy : 0.3050
##
##         'Positive' Class : Y
##
```

This result was worse than the logistic regression model.

Tree

The tree model is evaluated by specifying regmethod = "rpart". The candidate tuning parameters are set to NULL, so that only the complexity is selected as the default setting.

```
treegrid = NULL
ptest212 = ptest(object=fit311, Z=Z, newdata=SB2, testZ=testZ,
regmethod = "rpart", metric="ROC",
param=treegrid)
ptest212$trainout
```

```
## CART
##
## 40 samples
## 20 predictors
##  2 classes: 'N', 'Y'
##
## No pre-processing
## Resampling: Cross-Validated (10 fold, repeated 5 times)
## Summary of sample sizes: 36, 36, 36, 36, 36, 36, ...
## Resampling results across tuning parameters:
##
##   cp      ROC     Sens   Spec   Accuracy   Kappa   AUC
##   0.000   0.875   0.85   0.90   0.875      0.75    NaN
##   0.425   0.875   0.85   0.90   0.875      0.75    NaN
##   0.850   0.560   0.90   0.22   0.560      0.12    NaN
##   Precision   Recall   F
##   0.9266667   0.85     0.8613333
##   0.9266667   0.85     0.8613333
##   0.5933333   0.90     0.6706667
##
## ROC was used to select the optimal model using
##   the largest value.
## The final value used for the model was cp = 0.425.
```

The tree model according to the chosen complexity calculates the confusion matrix and the prediction accuracy metric in the test data.

```
ptest212$predcnfmat
```

```
## Confusion Matrix and Statistics
##
##               Reference
## Prediction   N   Y
```

```
##           N 415 147
##           Y  85 353
##
##                 Accuracy : 0.768
##                   95% CI : (0.7406, 0.7938)
##      No Information Rate : 0.5
##      P-Value [Acc > NIR] : < 2.2e-16
##
##                    Kappa : 0.536
##
##   Mcnemar's Test P-Value : 6.206e-05
##
##              Sensitivity : 0.7060
##              Specificity : 0.8300
##           Pos Pred Value : 0.8059
##           Neg Pred Value : 0.7384
##                Precision : 0.8059
##                   Recall : 0.7060
##                       F1 : 0.7527
##               Prevalence : 0.5000
##           Detection Rate : 0.3530
##     Detection Prevalence : 0.4380
##        Balanced Accuracy : 0.7680
##
##          'Positive' Class : Y
##
```

The results were similar to those of SVM.

Random Forest

The random forest is evaluated by specifying regmethod = "rf". The candidate tuning parameters are set to NULL, so that only the mtry is selected as the default setting.

```
rfgrid  =  NULL
ptest213 = ptest(object=fit311, Z=Z, newdata=SB2, testZ=testZ,
regmethod = "rf", metric="ROC",
param=rfgrid)
ptest213$trainout
```

```
## Random Forest
##
```

```
## 40 samples
## 20 predictors
##  2 classes: 'N', 'Y'
##
## No pre-processing
## Resampling: Cross-Validated (10 fold, repeated 5 times)
## Summary of sample sizes: 36, 36, 36, 36, 36, 36, ...
## Resampling results across tuning parameters:
##
##   mtry  ROC     Sens  Spec  Accuracy  Kappa  AUC
##    2    0.8525  0.80  0.78  0.790     0.58   NaN
##   11    0.9400  0.88  0.87  0.875     0.75   NaN
##   20    0.9450  0.87  0.88  0.875     0.75   NaN
##   Precision  Recall  F
##   0.8366667  0.80    0.7965986
##   0.9066667  0.88    0.8693333
##   0.9166667  0.87    0.8660000
##
## ROC was used to select the optimal model using
##   the largest value.
## The final value used for the model was mtry = 20.
```

The random forests model according to the chosen mtry calculates the confusion matrix and the prediction accuracy metric in the test data.

```
ptest213$predcnfmat
```

```
## Confusion Matrix and Statistics
##
##            Reference
## Prediction   N    Y
##          N  415  152
##          Y   85  348
##
##                 Accuracy : 0.763
##                   95% CI : (0.7354, 0.7891)
##     No Information Rate : 0.5
##     P-Value [Acc > NIR] : < 2.2e-16
##
##                    Kappa : 0.526
##
##   Mcnemar's Test P-Value : 1.81e-05
##
##              Sensitivity : 0.6960
```

```
##                    Specificity : 0.8300
##                 Pos Pred Value : 0.8037
##                 Neg Pred Value : 0.7319
##                      Precision : 0.8037
##                         Recall : 0.6960
##                             F1 : 0.7460
##                     Prevalence : 0.5000
##                 Detection Rate : 0.3480
##           Detection Prevalence : 0.4330
##              Balanced Accuracy : 0.7630
##
##               'Positive' Class : Y
##
```

The results were similar to those of tree model.

Deep learning

Several ML methods have been applied in brain image analysis to extract predictive classification models (Mateos-Pérez et al., 2018, Pellegrini et al., 2018). These help to automate clinical diagnosis by extracting high-dimensional beneficial features, such as the classification of AD and MCI patients using structural MRI. Of these methods, the most frequently used is the support vector machine (SVM), which estimates the optimal hyperplane that optimally separates disease and healthy groups (Rathore et al., 2017). Recently, however, its popularity has been declining. The reason for this is that SVM does not work well with raw data and the result depends on "selection of features," that is, the selection of the amount calculated from the original data used as the input for classification. Many selections are done manually or semi-automatically, which is time-consuming and subject to variability between evaluators, or require complex image preprocessing and time-consuming calculations (Vieira et al., 2017).

In contrast, deep learning (DL), which has developed from ML, has produced good results in each field (LeCun et al., 2015). Image analysis requires less image pre-processing and can automatically estimate the prediction function from raw data without the need for prior feature selection, making it more objective and less biased. Therefore, it is considered superior to conventional machine learning methods and has been applied to many classification problems between AD sufferers and healthy subjects (Vieira et al., 2017, Suk et al., 2017). The DL method mentioned there was used to further learn multiple prediction models using DL with a sparse automatic encoder, a 2D convolutional neural network (CNN), a 3D CNN and sparse regression analysis. The brain image is a single structural image. Although the AD classification accuracy of structural images was in the range of 86–93%, better results were obtained with DL (Basaia et al., 2019). It is also applied to treating other brain diseases (Vieira et al., 2017; Heinsfeld et al., 2018).

R example

Since the `ptest` function depends on the `caret` package, it can specify the regression model which is available in the `caret` package. One of them is the `mxnet` function which is the deep learning.

Firstly, the candidate parameters of the deep learning process are prepared.

```
layers0 = c(1, 5, 10); layers1 = c(0, 1, 5, 10)
rate0 = c(0, 0.25, 0.5, 0.75)
activation=c("relu", "sigmoid", "tanh", "softrelu")
mxnet.params = expand.grid(layer1=layers0, layer2=layers1,
layer3=0, learning.rate=0.1, momentum=0.9, dropout=0,
activation=activation[3])
```

The deep learning model is implemented by selecting the parameter as follows.

```
ptest215 = ptest(object=fit113, Z=Z, newdata=SB2, testZ=testZ,
regmethod = "mxnet", metric="Accuracy",
param=mxnet.params)
ptest215$trainout
```

```
## Neural Network
##
## 40 samples
##  2 predictor
##  2 classes: 'N', 'Y'
##
## No pre-processing
## Resampling: Cross-Validated (10 fold, repeated 5 times)
## Summary of sample sizes: 36, 36, 36, 36, 36, 36, ...
## Resampling results across tuning parameters:
##
```

layer1	layer2	ROC	Sens	Spec	Accuracy	Kappa
1	0	0.995	0.90	1.00	0.950	0.90
1	1	0.990	0.90	1.00	0.950	0.90
1	5	0.990	0.89	0.95	0.920	0.84
1	10	0.990	0.90	1.00	0.950	0.90
5	0	0.995	0.90	1.00	0.950	0.90
5	1	0.990	0.90	1.00	0.950	0.90
5	5	0.985	0.90	1.00	0.950	0.90
5	10	0.990	0.91	1.00	0.955	0.91

```
## 10        0    0.990 0.95 0.94 0.945    0.89
## 10        1    0.990 0.92 0.95 0.935    0.87
## 10        5    0.990 0.93 0.95 0.940    0.88
## 10       10    0.995 0.91 0.95 0.930    0.86
## AUC  Precision  Recall  F
## NaN  1.0000000   0.90   0.9333333
## NaN  1.0000000   0.90   0.9333333
## NaN  0.9633333   0.89   0.9073333
## NaN  1.0000000   0.90   0.9333333
## NaN  1.0000000   0.90   0.9333333
## NaN  1.0000000   0.90   0.9333333
## NaN  1.0000000   0.90   0.9333333
## NaN  1.0000000   0.91   0.9400000
## NaN  0.9600000   0.95   0.9446667
## NaN  0.9633333   0.92   0.9273333
## NaN  0.9633333   0.93   0.9340000
## NaN  0.9633333   0.91   0.9206667
##
## Tuning parameter 'layer3' was held constant at a
##  parameter 'activation' was held constant at a
##  value of tanh
## Accuracy was used to select the optimal model
##  using the largest value.
## The final values used for the model were layer1 =
##  0.1, momentum = 0.9, dropout = 0 and activation
##  = tanh.
```

The prediction performance is evaluated on the test data.

ptest215$predcnfmat

```
## Confusion Matrix and Statistics
##
##              Reference
## Prediction   N    Y
##          N  406   97
##          Y   94  403
##
##                   Accuracy : 0.809
##                     95% CI : (0.7832, 0.8329)
##        No Information Rate : 0.5
##        P-Value [Acc > NIR] : <2e-16
##
##                      Kappa : 0.618
##
##   Mcnemar's Test P-Value : 0.8849
```

```
##
##              Sensitivity : 0.8060
##              Specificity : 0.8120
##           Pos Pred Value : 0.8109
##           Neg Pred Value : 0.8072
##                Precision : 0.8109
##                   Recall : 0.8060
##                       F1 : 0.8084
##               Prevalence : 0.5000
##           Detection Rate : 0.4030
##     Detection Prevalence : 0.4970
##         Balanced Accuracy : 0.8090
##
##         'Positive' Class : Y
##
```

The test accuracy (0.809) was slightly better than the logistic regression model (0.806).

Summary

Brain image data is classified into ultra-high-dimensional data and it becomes more highly dimensional because of clinical necessity. Therefore, the step of reducing dimensionality is considered essential and it is necessary to use the data structure to interpret the result. Furthermore, a data-driven method is necessary to reduce the bias caused by the manual method. Matrix decomposition methods represented by principal component analysis are data-driven and extensions such as sparse estimation, supervised estimation and block modeling are useful for brain image analysis (Kawaguchi and Yamashita, 2017). Furthermore, there are many variations such as robust, probabilistic and deep tensor decomposition. Considering these with matrix decomposition, there is the possibility of obtaining important knowledge in brain image analysis. For the clinical application of image data as a biomarker, it must be certified by regulatory authorities and then be applied to various datasets for validation (Frisoni et al., 2017). Applications for machine learning and deep learning are also advancing in predictive models, but dimensional reduction by the matrix decomposition method and combined use with the two-steps dimensional reduction method are also useful. Additionally, considering the changes in the application of prediction models in clinical research, it is held that method imbalance and result invariance is needed. Therefore, it is necessary to develop a methodology that can assert this.

Chapter 5

Advance Methods

Multimodal analysis

So far, we have considered the brain image data matrix X obtained from a single measuring device, but now we will extend it to have multiple brain image data per person. By analyzing multiple data sets that have been analyzed individually, disease characterization is performed simultaneously from various angles. In brain image analysis, this is known as multimodality and it evaluates brain pathology from the perspectives of brain morphology and function. In bioinformatics, it is called multi-omics and comprehensively analyzes biomolecular information such as genome, transcription, protein, metabolism, etc.

Multi-block approach

A score that is representative of each modality is computed, and the score that it represents is also synthesized. A principal component analysis could be applied to each modality and further principal component analysis could be performed on the principal component scores obtained therein. Here, we introduce multi-block principal component analysis, which is a method developed from principal component analysis to integrate such multiple data sets, rather than a method to score them independently.

Assuming that there are M modalities, the $n \times p_m$ data matrix is $\boldsymbol{X}_m(m = 1, 2, \ldots, M)$. Principal component analysis is extended to become multi-block principal component analysis. The multi-block score (component) is

$$\boldsymbol{s} = \sum_{m=1}^{M} b_m \boldsymbol{X}_m \boldsymbol{w}_m$$

$\boldsymbol{w} = (\boldsymbol{w}_1, \boldsymbol{w}_2, \ldots, \boldsymbol{w}_M)^\top$ is called a sub-weight vector, $\boldsymbol{b} = (b_1, b_2, \ldots, b_M)^\top$ is called a super-weight vector. This is a method to obtain the weights within a data set (within each modality) and between them (between modalities) simultaneously, as shown in Figure 5.1. In particular, the between-modality weights represent the contribution of the modality and can also be important in prediction models and may be useful in selecting the modality to be measured according to its degree.

Each weight maximizes the variance of the multi-block score $\boldsymbol{s}$ under the constraints $\|\boldsymbol{w}_m\|^2 = 1$ and $\|\boldsymbol{b}\|^2 = 1$. In other words, the following function is maximized.

$$L(\boldsymbol{b}, \boldsymbol{w}) = \boldsymbol{s}^\top \boldsymbol{s} - \sum_{m=1}^{M} \eta_{1,m} \boldsymbol{w}_m^\top \boldsymbol{w}_m - \eta_2 \boldsymbol{b}^\top \boldsymbol{b}$$

where $\eta_{1,m} > 0$, $\eta_2 > 0$ are multipliers.

Performing regression analysis between matrices is also applied and PLS and CCA correspond to this framework. Here we consider two modalities. Let $\boldsymbol{X}$ be the data matrix in the first modality of $n \times p$, and $\boldsymbol{Y}$ be the data matrix in the

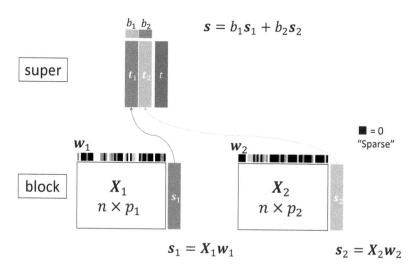

Figure 5.1: Multi-block scoring approach.

second modality of $n \times q$. Assume that each column is standardized to mean 0 and variance 1. The purpose of the analysis is to find a score for the relationship between X and Y. The scores t and u for X and Y respectively are given as follows.

$$t = X w_X, \quad u = Y w_Y$$

w_X is p-dimensional, w_Y is q-dimensional weight vector, obtained by maximizing correlation between scores and scores and outcomes. In other words, the optimization problem $\max L_0(w_X, w_Y)$ is considered where

$$L_0(w_X, w_Y) = t^T u - \eta_X w_X^\top w_X - \eta_Y w_Y^\top w_Y$$

with $\eta_X > 0$ and $\eta_Y > 0$.

This makes it possible to use high-dimensional data such as laboratory data (Yoshida et al., 2018) or genetic data (Imaging genetics, Nathoo et al., 2019) as explanatory variables, with images as objective variables. Thus, instead of clinical information, image data can be an objective variable, so it is also called an intermediate phenotype.

MSMA method

In this section, we present the multi-block sparse multivariate analysis (MSMA) method proposed by Kawaguchi and Yamashita (2017) as a general framework of methods introduced in the previous sections. Consider n subjects and M_X and M_Y blocks (modalities) for the predictor and response, respectively. We assume that modalities are given. X_m is the $n \times p_m$ ($m = 1, 2, \ldots, M_X$) and Y_m is the $n \times q_m$ ($m = 1, 2, \ldots, M_Y$) for the m-th block (modality) predictor and response matrix, respectively. Each block corresponds to the modality, for example, Y_1 is MRI, Y_2 is PET and X_m is SNP. Subjects also have a univariate outcome, and the n-dimensional outcome vec-tor is denoted by Z. We consider scores t and u for the predictor and response, respectively, with the following multi-block structure.

$$t = \sum_{m=1}^{M_X} b_{X,m} X_m w_{X,m}, \quad u = \sum_{m=1}^{M_Y} b_{Y,m} Y_m w_{Y,m} \tag{5.1}$$

where $w_{X,m}$ and $w_{Y,m}$ are weight vectors for the m-th sub-blocks X_m and Y_m, respectively, and $b_{X,m}$ and $b_{Y,m}$ are the weights for the super-block. Here it should be noted that the scores in (5.1) are referred to as the super score, whereas $t_m = X_m w_{X,m}$ and $u_m = Y_m w_{Y,m}$ are referred to as the block score. Thus, the super score, which is used in an application such as a prediction, has a hieratical structure.

This scoring has a low dimensional representation of not only the data in each block but also the relationship between blocks. Thus, this would be useful for modeling statistically such as construction of the prediction model as in our application by using the multimodality of images and genetics whose importance is discussed in the previous section. We could also evaluate the effects of modalities through $b_{X,m}$ and $b_{Y,m}$. From an application perspective in our case, it is necessary to detect the pair of images and genomic components with the block feature, in which the image is used as marker. Our aim is to determine the associated genomic information with the marker. Thus, the association between two data sets (for example, images and SNPs) is evaluated by maximizing the correlation between the super scores t and u supervised by the outcome. Actually, when matrices X_m and Y_m are normalized by their columns, the weights $w_X = (w_{X,1}, w_{X,2}, \ldots, w_{X,M_X})^\top$, $w_Y = (w_{Y,1}, w_{Y,2}, \ldots, w_{Y,M_Y})^\top$, $b_X = (b_{X,1}, b_{X,2}, \ldots, b_{X,M_X})^\top$ and $b_Y = (b_{Y,1}, b_{Y,2}, \ldots, b_{Y,M_Y})^\top$ are estimated by maximizing the following function.

$$L(w_X, w_Y, b_X, b_Y) = \mu_{XY} t^\top u + \mu_{XZ} t^\top Z + \mu_{YZ} u^\top Z$$
$$- \sum_{m=1}^{M_X} P_{\lambda_{X,m}}(w_{X,m}) - \sum_{m=1}^{M_Y} P_{\lambda_{Y,m}}(w_{Y,m}) \qquad (5.2)$$

subject to $\|w_{X,m}\|_2^2 = 1$, $\|w_{Y,m}\|_2^2 = 1$, for any m, $\|b_X\|_2^2 = 1$, $\|b_Y\|_2^2 = 1$, where $0 \le \mu_{XY}, \mu_{XZ}, \mu_{YZ} \le 1$ and $\mu_{XY} + \mu_{XZ} + \mu_{YZ} = 1$, is the proportion of the supervision, $P_\lambda(x)$ is the penalty function, $(P_\lambda(x) = 2\lambda|x|$ is used in the application of this paper because of the simplicity and the possibility for the sparsity of the weights), and $\lambda > 0$ is the regularized parameter that is used to control the sparsity. Note that it would be possible for the penalty function to have other structures, as in the previous section.

Note that when $\mu_{XZ} = 0$ and $\mu_{YZ} = 0$ the regularized multi-block PLS or CCA is induced, and when $X_m = Y_m$ the supervised regularized PCA framework is induced. These could fit flexibly to several applications according to clinical interests. The proposed method is closely related to that of Li et al.(2012), which is the unsupervised version and was applied to multi-genomic data analysis without imaging data.

We estimate the weights in (5.1) by stating the following proposition.

Proposition 1 *The solution of optimization problem (5.2) satisfies*

$$\tilde{b}_{X,m} = t_m^\top \{\mu_{XY} u + \mu_{XZ} Z\}, \quad \tilde{b}_{Y,m} = \{\mu_{XY} t^\top + \mu_{YZ} Z^\top\} u_m$$
$$\tilde{w}_{X,m} = h_{\lambda_{X,m}}(b_{X,m} X_m^\top \{\mu_{XY} u + \mu_{XZ} Z\}),$$
$$\tilde{w}_{Y,m} = h_{\lambda_{Y,m}}(b_{Y,m} \{\mu_{XY} t^\top + \mu_{YZ} Z^\top\} Y_m)$$

where $h_\lambda(y) = sign(y)(|y| - \lambda)_+$ is the sparse function. The proof is shown later.

This leads to the following algorithm to estimate the weights in (5.1) by maximizing L in (5.2).

1. Initialize t and u and normalize the super scores as follows.

$$t \leftarrow t/\|t\|_2, \quad u \leftarrow u/\|u\|_2$$

where $\leftarrow$ means "replaced with".

2. Repeat until convergence.

   (a) For fixed u, $\tilde{w}_{X,m} = h_{\lambda_{X,m}}(b_{X,m}X_m^\top\{\mu_{XY}u + \mu_{XZ}Z\})$ and normalize $\hat{w}_{X,m} = \tilde{w}_{X,m}/\|\tilde{w}_{X,m}\|_2$ $(m = 1,2,\ldots,M_X)$.

   (b) Putting $t_m = X_m\hat{w}_{X,m}$, $\tilde{b}_{X,m} = t_m^\top\{\mu_{XY}u + \mu_{XZ}Z\}$ then putting $\tilde{b}_X = (\tilde{b}_{X,1}, \tilde{b}_{X,2}, \ldots, \tilde{b}_{X,M_X})^\top$ and normalize as $\hat{b}_X = \tilde{b}_X/\|\tilde{b}_X\|_2$

   (c) For $t = \sum_{m=1}^{M_X} b_{X,m}X_m\hat{w}_{X,m}$, $\tilde{w}_{Y,m} = h_{\lambda_{Y,m}}(b_{Y,m}\{\mu_{XY}t^\top + \mu_{YZ}Z^\top\}Y_m)$ then normalize $\hat{w}_{Y,m} = \tilde{w}_{Y,m}/\|\tilde{w}_{Y,m}\|_2$ $(m = 1,2,\ldots,M_Y)$.

   (d) Putting $u_m = Y_m\hat{w}_{Y,m}$, $\tilde{b}_{Y,m} = \{\mu_{XY}t^\top + \mu_{YZ}Z^\top\}u_m$ then putting $\tilde{b}_Y = (\tilde{b}_{Y,1}, \tilde{b}_{Y,2}, \ldots, \tilde{b}_{Y,M_Y})^\top$ and normalize as $\hat{b}_Y = \tilde{b}_Y/\|\tilde{b}_Y\|_2$.

   (e) set $u = \sum_{m=1}^{M_Y} \hat{b}_{Y,m}u_m$ where $\hat{b}_{Y,m}$ is the m-the element of $\hat{b}_Y$.

3. Set $p_m = X_m^\top t_m/t_m^\top t_m$ and $q_m = Y_m^\top u_m/u_m^\top u_m$, $X_m \leftarrow X_m - t_m p_m^\top$ and $Y_m \leftarrow Y_m - u_m q_m^\top$.

Note that the deflation step yields multiple components and has several alternatives. The convergence of this algorithm was supported by the proposition provided in the supplementary materials. The larger value of the regularization parameter $\lambda_{X,m}$ or $\lambda_{Y,m}$ has many non-zero elements in $w_{X,m}$ or $w_{Y,m}$.

The method for the regularization parameters selection was also provided in following. As in Shen and Huang (2008), the characteristics for scores can be considered. First, we consider the projection of Y_m onto the k-dimensional subspace spanned by the k loading vectors as $\hat{Y}_m^{(k)} = U_m Q_m^\top$, where $U_m = [u_m^{(1)},\ldots,u_m^{(k)}]$ and $Q_m = [q_m^{(1)},\ldots,q_m^{(k)}]$. Define the adjusted variance of the k-th score as

$$\mathrm{tr}(\hat{Y}_m^{(k)\top}\hat{Y}_m^{(k)}) - \mathrm{tr}(\hat{Y}_{m-1}^{(k)\top}\hat{Y}_{m-1}^{(k)}).$$

We also define the cumulative percentage of the explained variance (CPEV) by the first k scores as

$$\mathrm{tr}(\hat{Y}_m^{(k)\top}\hat{Y}_m^{(k)})/\mathrm{tr}(Y_m^\top\hat{Y}_m).$$

These definitions can be applied to every m and X_m.

The larger value of the regularization parameter $\lambda_{X,m}$ or $\lambda_{Y,m}$ has many non-zero elements in $w_{X,m}$ or $w_{Y,m}$, values from which its optimal value is selected by minimizing the Bayesian information criterion (BIC) that was proposed by Lee et al. (2010) and Allen et al. (2014).

$$
\text{BIC}(\lambda_X, \lambda_Y) = \log\left(\frac{\sum_{m=1}^{k} \|\hat{\mathbf{Y}}_{m-1}^{(k)} - \mathbf{Y}_m\|^2}{n\sum_{m=1}^{k} q_m}\right) + \frac{\log(n\sum_{m=1}^{k} q_m)}{n\sum_{m=1}^{k} q_m} \text{df}(\lambda_X, \lambda_Y),
$$

where $\boldsymbol{\lambda}_X = (\lambda_{X,1}, \ldots, \lambda_{X,M_X})^\top$, $\boldsymbol{\lambda}_Y = (\lambda_{Y,1}, \ldots, \lambda_{Y,M_Y})^\top$, and df is the number of effective parameters, which depend on the value of $\boldsymbol{\lambda}_X$ and $\boldsymbol{\lambda}_Y$. Many workers have proposed cross-validation approaches for the SVD (Owen and Perry, 2009), but they would be computationally expensive.

The proof for the proposition is provided.

Proof of Proposition 1

First, consider equation (5.2) as the optimization problem $\max L_0(\mathbf{w}_X, \mathbf{w}_Y, \mathbf{b}_X, \mathbf{b}_Y)$ subject to $\|\mathbf{w}_{X,m}\|_2^2 = 1$, $\|\mathbf{w}_{Y,m}\|_2^2 = 1$, for any m, $\|\mathbf{b}_X\|_2^2 = 1$, $\|\mathbf{b}_Y\|_2^2 = 1$, where

$$
L_0(\mathbf{w}_X, \mathbf{w}_Y, \mathbf{b}_X, \mathbf{b}_Y) = \mu_{XY}\mathbf{t}^\top\mathbf{u} + \mu_{XZ}\mathbf{t}^\top\mathbf{Z} + \mu_{YZ}\mathbf{u}^\top\mathbf{Z} - \sum_{i=1}^{M_X} P_{\lambda_{X,i}}(\mathbf{w}_{X,i}) - \sum_{j=1}^{M_Y} P_{\lambda_{Y,j}}(\mathbf{w}_{Y,j})
$$

with $0 \le \mu_{XY}, \mu_{XZ}, \mu_{YZ} \le 1$ and $\mu_{XY} + \mu_{XZ} + \mu_{YZ} = 1$, $P_\lambda(x) = 2\lambda|x|$. From (5.1),

$$
\mathbf{t}^\top\mathbf{u} = \sum_{i=1}^{M_X}\sum_{j=1}^{M_Y} b_{Xi}b_{Yj}\mathbf{t}_i^\top\mathbf{u}_j, \quad \mathbf{t}^\top\mathbf{Z} = \sum_{i=1}^{M_X} b_{Xi}\mathbf{t}_i^\top\mathbf{Z}, \quad \mathbf{u}^\top\mathbf{Z} = \sum_{j=1}^{M_Y} b_{Yj}\mathbf{u}_j^\top\mathbf{Z}.
$$

Taking into account the constraints, Lagrangian optimization is applied, that is, consider $\max L(\mathbf{w}_X, \mathbf{w}_Y, \mathbf{b}_X, \mathbf{b}_Y)$, where

$$
\begin{aligned}
L(\mathbf{w}_X, \mathbf{w}_Y, \mathbf{b}_X, \mathbf{b}_Y) = {} & L_0(\mathbf{w}_X, \mathbf{w}_Y, \mathbf{b}_X, \mathbf{b}_Y) \\
& - \sum_{i=1}^{M_X} \eta_{1i}\|\mathbf{w}_{X,i}\|_2^2 - \sum_{j=1}^{M_Y} \eta_{2j}\|\mathbf{w}_{Y,j}\|_2^2 - \eta_3\|\mathbf{b}_X\|_2^2 - \eta_4\|\mathbf{b}_Y\|_2^2
\end{aligned}
$$

with $\eta_{1i} > 0$, $\eta_{2j} > 0$, $\eta_3 > 0$, $\eta_4 > 0$. Estimates of the super weight for X are derived from the derivative

$$
\frac{\partial L(\mathbf{w}_X, \mathbf{w}_Y, \mathbf{b}_X, \mathbf{b}_Y)}{\partial b_{Xi}} = \mu_{XY}\sum_{j=1}^{M_Y} b_{Yj}\mathbf{t}_i^\top\mathbf{u}_j + \mu_{XZ}\mathbf{t}_i^\top\mathbf{Z} - 2\eta_3 b_{Xi}
$$

by solving the equation $\partial L(\mathbf{w}_X, \mathbf{w}_Y, \mathbf{b}_X, \mathbf{b}_Y)/\partial b_{Xi} = 0$. Thus, the solution is given by

$$
\tilde{b}_{Xi} = \frac{1}{2\eta_3}\mathbf{t}_i^\top\{\mu_{XY}\mathbf{u} + \mu_{XZ}\mathbf{Z}\}.
$$

The final estimate is obtained by normalizing $\hat{\boldsymbol{b}}_X = \tilde{\boldsymbol{b}}_X / \|\tilde{\boldsymbol{b}}_X\|_2$ where $\tilde{\boldsymbol{b}}_X = (\tilde{b}_{X,1},$ $\ldots, \tilde{b}_{X,M_X})^\top$. Similarly, estimates of the super weight for Y are derived from the derivative

$$\frac{\partial L(\boldsymbol{w}_X, \boldsymbol{w}_Y, \boldsymbol{b}_X, \boldsymbol{b}_Y)}{\partial b_{Yj}} = \mu_{XY} \sum_{i=1}^{M_X} b_{Yj} t_i^\top \boldsymbol{u}_j + \mu_{YZ} \boldsymbol{u}^\top \boldsymbol{Z} - 2\eta_4 b_{Yj}$$

by solving the equation $\partial L(\boldsymbol{w}_X, \boldsymbol{w}_Y, \boldsymbol{b}_X, \boldsymbol{b}_Y)/\partial b_{Yj} = 0$. Thus, the solution is given by

$$\tilde{b}_{Yj} = \frac{1}{2\eta_4} \{\mu_{XY} t^\top + \mu_{XZ} \boldsymbol{Z}^\top\} \boldsymbol{u}_j.$$

The final estimate is obtained by normalizing $\hat{\boldsymbol{b}}_Y = \tilde{\boldsymbol{b}}_Y / \|\tilde{\boldsymbol{b}}_Y\|_2$ where $\tilde{\boldsymbol{b}}_Y = (\tilde{b}_{Y,1},$ $\ldots, \tilde{b}_{Y,M_X})^\top$.

A block weight for X is obtained by considering the following objective function.

$$
\begin{aligned}
L_1(\boldsymbol{w}_X) \;:=\; & \mu_{XY} \sum_{i=1}^{M_X} \sum_{j=1}^{M_Y} b_{Xi} b_{Yj} \boldsymbol{w}_{Xi}^\top \boldsymbol{X}_i^\top \boldsymbol{u}_j + \mu_{XZ} \sum_{i=1}^{M_X} b_{Xi} \boldsymbol{w}_{Xi}^\top \boldsymbol{X}_i^\top \boldsymbol{Z} \\
& - \sum_{i=1}^{M_X} P_{\lambda_{Xi}}(\boldsymbol{w}_{Xi}) - \sum_{i=1}^{M_X} \eta_{1i} \|\boldsymbol{w}_{X,i}\|_2^2 \\
=\; & \sum_{k} \sum_{i=1}^{M_X} \left\{ (\mu_{XY} b_{Xi} \boldsymbol{X}_i^\top \boldsymbol{u} + \mu_{XZ} b_{Xi} \boldsymbol{X}_i^\top \boldsymbol{Z})_k w_{Xik} \right\} \\
& - \sum_{k} \sum_{i=1}^{M_X} \left\{ P_{\lambda_{Xi}}(w_{Xik}) + \eta_{1i} w_{Xik}^2 \right\}.
\end{aligned}
$$

From the Lemma,

$$\arg\max L_1(\boldsymbol{w}_X) = \frac{1}{2\eta_{1i}} h_{\lambda_{Xi}} \left(b_{Xi} \boldsymbol{X}_i^\top \{\mu_{XY} \boldsymbol{u} + \mu_{XZ} \boldsymbol{Z}\} \right) =: \tilde{w}_{Xi}.$$

Thus, the final estimate is obtained by normalizing $\hat{\boldsymbol{w}}_X = \tilde{\boldsymbol{w}}_X / \|\tilde{\boldsymbol{w}}_X\|_2$ where $\tilde{\boldsymbol{w}}_X = (\tilde{w}_{X,1}, \ldots, \tilde{w}_{X,M_X})^\top$. Similarly, the block weight for Y is obtained by considering the following objective function.

$$
\begin{aligned}
L_2(\boldsymbol{w}_Y) \;:=\; & \mu_{XY} \sum_{i=1}^{M_X} \sum_{j=1}^{M_Y} b_{Xi} b_{Yj} t_i^\top \boldsymbol{Y}_j \boldsymbol{w}_{Yj} + \mu_{YZ} \sum_{j=1}^{M_Y} b_{Yj} \boldsymbol{Z}^\top \boldsymbol{Y}_j \boldsymbol{w}_{Yj} \\
& - \sum_{j=1}^{M_Y} P_{\lambda_{Yj}}(\boldsymbol{w}_{Yj}) - \sum_{j=1}^{M_Y} \eta_{2j} \|\boldsymbol{w}_{Yj}\|_2^2 \\
=\; & \sum_{k} \sum_{j=1}^{M_Y} \left\{ (\mu_{XY} b_{Yj} t^\top \boldsymbol{Y}_j + \mu_{YZ} b_{Yj} \boldsymbol{Z}^\top \boldsymbol{Y}_j)_k w_{Yjk} \right\} \\
& - \sum_{k} \sum_{j=1}^{M_Y} \left\{ P_{\lambda_{Yj}}(w_{Yjk}) + \eta_{2j} w_{Yjk}^2 \right\}.
\end{aligned}
$$

From the Lemma,

$$\text{argmax } L_2(\mathbf{w}_Y) = \frac{1}{2\eta_{2j}} h_{\lambda_{Yj}} \left(b_{Yj} \left\{ \mu_{XY} \mathbf{t}^\top + \mu_{YZ} \mathbf{Z}^\top \right\} \mathbf{Y}_j \right) =: \tilde{w}_{Yj}.$$

Thus, the final estimate is obtained by normalizing $\hat{\mathbf{w}}_Y = \tilde{\mathbf{w}}_Y / \|\tilde{\mathbf{w}}_Y\|_2$ where $\tilde{\mathbf{w}}_Y = (\tilde{w}_{Y,1}, \dots, \tilde{w}_{Y,M_Y})^\top$. The proof is completed.

Tensor decomposition

Since multimodal data has multiple data matrices, it can be considered as a data tensor by combining them into an array and tensor decomposition can also be applied. There are two types of tensor decomposition, CP (canonical polyadic) and Tucker, but Tucker is more generalized. Let $\underline{X}$ be $n \times N \times M$ data tensorD Each axis of the tensor is called a mode and its length is called a dimension. Here, mode 1 represents an n-dimensional subject, mode 2 represents an N-dimensional spatial domain (voxel) and mode 3 represents an M-dimensional modality. Then the tensor decomposition is given by

$$\underline{X} = C \times_1 S \times_2 W_1 \times_3 W_2$$

where the $K_1 \times K_2 \times K_3$ tensor C is called a core tensor. Each dimension of this core tensor can be viewed as the number of components for each mode of the data tensor. In this way Tucker alters the number of components in each mode. S is $n \times K_1$ score matrix, W_1 is the $N \times K_1$ weight matrix for spatial domain, and W_2 is the $M \times K_3$ weight matrix for modalities. $\times_\ell$ is the mode ℓ product (summed for mode ℓ). The ijm-component is as follows.

$$X_{ijm} = \sum_{k_1=1}^{K_1} \sum_{k_2=1}^{K_2} \sum_{k_3=1}^{K_3} c_{k_1 k_2 k_3} S_{ik_1} W_{1jk_2} W_{2mk_3}.$$

The objective function is $L = \|\underline{X} - C \times_1 S \times_2 W_1 \times_3 W_2\|_F^2$, but C that minimizes this becomes $C = \underline{X} \times_2 W_1 \times_3 W_2$, substituting into L and deforming results in a problem where $L_2 = \|C \times_1 S \times_2 W_1 \times_3 W_2\|_F^2$ needs to be maximized. If this is rewritten with ingredients

$$L_2 = \sum_{k_1=1}^{K_1} \sum_{k_2=1}^{K_2} \sum_{k_3=1}^{K_3} \left(\sum_{i=1}^{n} \sum_{j=1}^{N} \sum_{m=1}^{M} X_{ijm} S_{ik_1} W_{1jk_2} W_{2mk_3} \right)^2$$

$$= \sum_{k_1=1}^{K_1} \sum_{k_2=1}^{K_2} \sum_{k_3=1}^{K_3} \left(\sum_{i=1}^{n} S_{ik_1} \sum_{j=1}^{N} \sum_{m=1}^{M} X_{ijm} W_{1jk_2} W_{2mk_3} \right)^2$$

and $L_2 = \|S^\top X_{(1)} (W_2 \otimes W_1)\|_F^2$ where $\otimes$ is the Kronecker product. This means that $X_{(1)}(W_2 \otimes W_1)$ is decomposed into singular values and the left singular vector is given as S, where $X_{(\ell)}$ is an ℓ-way matrix and $X_{(1)}$ is an $n \times NM$ matrix.

Similarly, W_1 is the left singular vector of $X_{(2)}(W_2 \otimes S)$ and W_2 is the left singular vector of $X_{(3)}(W_1 \otimes S)$. Multi-view learning and multi-task learning are examples of learning methods related to multi-modal analysis.

R example

Multi-block PCA

Generate simulation data

The data is generated by the `strsimdata` function. The function generates data by applying a zero-weighted load to randomly generated factors.

```
n = 100; seed = 2
dataset1 = strsimdata(n = n, ncomp=2,
Xps=c(4,4), Ztype="binary", seed=seed)
```

The number of subjects is 100, the number of factors is 2, the generated explanatory variable matrix X has 2 blocks and the number of variables is 4 and 4, respectively. Thus, the number of blocks is set by the length of the vector that specifies the number of variables. Also, set whether to generate the supervisor vector Z. Multi-block data is a list of data matrices.

```
X2 = dataset1$X;
Z = dataset1$Z
str(dataset1[c("X","Z")])

## List of 2
##  $ X:List of 2
##   ..$ : num [1:100, 1:4] -280.05 -154.23 77.88 193.64
     4.69 ...
##   .. ..- attr(*, "dimnames")=List of 2
##   .. .. ..$ : NULL
##   .. .. ..$ : chr [1:4] "v1" "v2" "v3" "v4"
##   ..$ : num [1:100, 1:4] -1.37e-14 -2.29e-15 5.99e-15
     2.88e-15 2.69e-15 ...
##   .. ..- attr(*, "dimnames")=List of 2
##   .. .. ..$ : NULL
##   .. .. ..$ : chr [1:4] "v1" "v2" "v3" "v4"
##  $ Z: int [1:100] 1 1 0 1 0 0 0 0 1 1 ...
```

The weights for X are generated by normalizing normal random numbers so that their length is 1 and they are stored as follows.

```
dataset1$WX

## $block
## $block$block1
##             comp1       comp2
## v1 -0.4164886   0.0000000
## v2  0.0000000  -0.4527662
## v3  0.7373273  -0.5818499
## v4 -0.5248980   0.6708455
##
## $block$block2
##             comp1       comp2
## v1  0.0000000   0.0000000
## v2  0.3646608  -0.5889031
## v3  0.8571916   0.0000000
## v4 -0.3428716   0.5919526
##
##
## $super
##               comp1       comp2
## block1 0.9197372   0.8673174
## block2 0.3925346  -0.4977554
```

The first element of the list has a super weight and the second element has a block weight. The block weight corresponds to a component in the row and the column corresponds to the number of variables.

The numbers of zero weights are as follows.

```
dataset1$nZeroX

## $block
## $block$block1
## [1] 1 1
##
## $block$block2
## [1] 1 2
##
##
## $super
## $super[[1]]
## NULL
##
## $super[[2]]
## NULL
```

```
dataset1$ZcoefX
```

```
## [1] 0.2977618 0.7081123 0.3182731 0.5555332
```

Supervised multi-block PCA

Perform supervised multi-block PCA by setting not only X2 but also Z and supervised parameter muX. First, select the number of components and the regularized parameter.

```
(opt212 = optparasearch(X2, Z=Z, muX=0.5,
search.method = "ncomp1st", criterion="BIC", whichselect="X"))
```

```
## Search method: ncomp1stSearch criterion: BIC
##
## Optimal number of components: 2
##
## Optimal parameters:
##
## lambdaX1 lambdaX2
##    0.355    0.462
```

Perform supervised multi-block PCA using msma function using the selected number of components and regularized parameters.

```
(fit212 = msma(X2, Z=Z, muX=0.5, comp=opt212$optncomp,
lambdaX=opt212$optlambdaX))
```

```
## Call:
## msma.default(X = X2, Z = Z, comp = opt212$optncomp,
    lambdaX = opt212$optlambdaX,
##     muX = 0.5)
##
## Numbers of non-zeros for X:
##        comp1 comp2
## block1     3     2
## block2     3     1
##
## Numbers of non-zeros for X super:
## comp1 comp2
##     2     2
```

The results of the first and the second components are as follows.

```
par(mfrow=c(2,2), oma = c(0, 0, 2, 0))
plot(fit212, axes = 1, plottype="bar",
block="super", XY="X")
plot(fit212, axes = 2, plottype="bar",
```

```
block="super", XY="X")
plot(fit212, axes = 1, plottype="bar",
block="block", XY="X")
plot(fit212, axes = 2, plottype="bar",
block="block", XY="X")
```

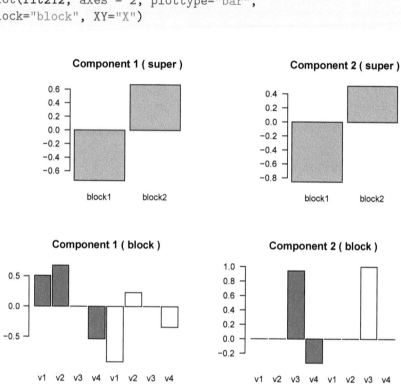

The relationship between the super score and the binary outcome Z is examined.

```
par(mfrow=c(1,2))
for(i in 1:2){
t1=t.test(fit212$ssX[,i]~Z)
boxplot(fit212$ssX[,i]~Z,
main=paste("Comp", i),
sub=paste("t-test p =", round(t1$p.value,4)))
}
```

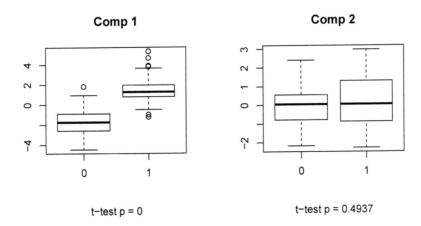

t-test p = 0   t-test p = 0.4937

Tensor decomposition

Load the rTensor package to perform tensor decomposition.

```
library(rTensor)
```

Convert data from list format to an array format and convert it to the tensor format.

```
(X3 = as.tensor(array(unlist(X2),
dim=c(dim(X2[[1]]),length(X2)))))
```

```
## Numeric Tensor of 3 Modes
## Modes:  100 4 2
## Data:
## [1] -280.052003 -154.229507   77.876078  193.641897
## [5]    4.693637    9.003032
```

This tensor has three modes. The first corresponds to the subject, the second corresponds to the variable and the third corresponds to the block.

First, the canonical polyadic (CP) decomposition is applied.

```
cpD = cp(X3, num_components=2)
```

```
##   |                                                    |
```

The decomposition output is stored in the matrix object U, where each column number corresponds to the number of components.

```
lapply(cpD$U, dim)
```

```
## [[1]]
```

```
## [1] 100    2
##
## [[2]]
## [1] 4 2
##
## [[3]]
## [1] 2 2
```

Next, the Tucker decomposition is applied.

```
tuckerD = tucker(X3, ranks=c(2,2,2))
```

```
##    |                                                              |
```

The core tensor has three modes with two components.

```
tuckerD$Z
```

```
## Numeric Tensor of 3 Modes
## Modes:  2 2 2
## Data:
## [1] 1970.45426    73.77168   -69.48489   774.65822
## [5]  -94.10101   243.11906
```

The decomposition output is the same as in the CP decomposition.

```
lapply(tuckerD$U, dim)
```

```
## [[1]]
## [1] 100    2
##
## [[2]]
## [1] 4 2
##
## [[3]]
## [1] 2 2
```

PLS

Generate simulation data

The data is generated by the `strsimdata` function. The function generates data by applying a zero-weighted load to randomly generated factors.

```
dataset2 = strsimdata(n = n, ncomp=2, Xps=c(4,4),
Yps=c(3,5), Ztype="binary", cz=c(10,10), seed=seed)
```

The number of subjects is 100, the number of factors is 2, the generated explanatory variable matrix X has 2 blocks and the number of variables is 4 and 5, respectively. Thus, the number of blocks is set by the length of the vector that specifies the number of variables. The same is true for the objective variable matrix Y. Also, set whether to generate the supervisor vector Z. It is generated here.

Multi-block data is a list of data matrices.

```
X2 = dataset2$X; Y2 = dataset2$Y
Z = dataset2$Z
str(dataset2[c("X","Y","Z")])
```

```
## List of 3
##  $ X:List of 2
##   ..$ : num [1:100, 1:4] -280.05 -154.23 77.88 193.64
      4.69 ...
##   .. ..- attr(*, "dimnames")=List of 2
##   .. .. ..$ : NULL
##   .. .. ..$ : chr [1:4] "v1" "v2" "v3" "v4"
##   ..$ : num [1:100, 1:4] -1.37e-14 -2.29e-15 5.99e-15
      2.88e-15 2.69e-15 ...
##   .. ..- attr(*, "dimnames")=List of 2
##   .. .. ..$ : NULL
##   .. .. ..$ : chr [1:4] "v1" "v2" "v3" "v4"
##  $ Y:List of 2
##   ..$ : num [1:100, 1:3] -9.925 -1.318 3.146 0.979
      -0.106 ...
##   .. ..- attr(*, "dimnames")=List of 2
##   .. .. ..$ : NULL
##   .. .. ..$ : chr [1:3] "v1" "v2" "v3"
##   ..$ : num [1:100, 1:5] 0.345 0.442 0.136 -0.68 0.035 ...
##   .. ..- attr(*, "dimnames")=List of 2
##   .. .. ..$ : NULL
##   .. .. ..$ : chr [1:5] "v1" "v2" "v3" "v4" ...
##  $ Z: int [1:100] 1 1 0 0 0 1 0 0 1 1 ...
```

The weights for X are generated by normalizing normal random numbers so that their length is 1 and they are stored as follows.

```
dataset2$WX
```

```
## $block
## $block$block1
##           comp1      comp2
## v1 -0.4164886  0.0000000
```

```
## v2   0.0000000 -0.4527662
## v3   0.7373273 -0.5818499
## v4  -0.5248980  0.6708455
##
## $block$block2
##             comp1       comp2
## v1   0.0000000  0.0000000
## v2   0.3646608 -0.5889031
## v3   0.8571916  0.0000000
## v4  -0.3428716  0.5919526
##
##
## $super
##              comp1        comp2
## block1 0.9197372  0.8673174
## block2 0.3925346 -0.4977554
```

The first element of the list has a super weight and the second element has a block weight. The block weight corresponds to a component in the row and the column corresponds to the number of variables.

The numbers of zero weights are as follows.

```
dataset2$nZeroX
```

```
## $block
## $block$block1
## [1] 1 1
##
## $block$block2
## [1] 1 2
##
##
## $super
## $super[[1]]
## NULL
##
## $super[[2]]
## NULL
```

The weights of Y as well as X are set as follows.

```
dataset2$WY
```

```
## $block
## $block$block1
```

```
##           comp1       comp2
## v1 -0.3809511  0.0000000
## v2  0.9239761 -0.6828194
## v3  0.0000000  0.0000000
##
## $block$block2
##           comp1       comp2
## v1  0.6580112  0.0000000
## v2  0.4583716  0.0000000
## v3 -0.5694518 -0.5602867
## v4 -0.1804586  0.0000000
## v5  0.0000000 -0.7604403
##
##
## $super
##              comp1      comp2
## block1 -0.02830175 0.9054941
## block2  0.99959943 0.4243588
```

dataset2$nZeroY

```
## $block
## $block$block1
## [1] 1 2
##
## $block$block2
## [1] 1 3
##
##
## $super
## $super[[1]]
## NULL
##
## $super[[2]]
## NULL
```

dataset2$ZcoefX

```
## [1] 3.015150 7.044889 3.170360 5.588133
```

dataset2$ZcoefY

```
## [1] 9.9689852 0.2838427 0.3100364 0.6653157
```

Supervised sparse PLS

Here, further set Z and execute supervised sparse PLS. The supervised parameters muX and muY are both set to 0.3 here.

```
(opt222 = optparasearch(X2, Y2, Z, muX=0.3, muY=0.3,
search.method = "ncomp1st", criterion="BIC",
criterion4ncomp="BIC", whichselect=c("X","Y")))
```

```
## Search method: ncomp1stSearch criterion: BIC
##
## Optimal number of components: 2
##
## Optimal parameters:
##
## lambdaX1 lambdaX2
##    0.143    0.393
##
## lambdaY1 lambdaY2
##    0.176    0.468
```

```
(fit222 = msma(X2, Y2, Z,
muX=0.3, muY=0.3, comp=opt222$optncomp,
lambdaX=opt222$optlambdaX, lambdaY=opt222$optlambdaY))
```

```
## Call:
## msma.default(X = X2, Y = Y2, Z = Z, comp = opt222$optncomp,
    lambdaX = opt222$optlambdaX,
##      lambdaY = opt222$optlambdaY, muX = 0.3, muY = 0.3)
##
## Numbers of non-zeros for X:
##        comp1 comp2
## block1    4     3
## block2    4     1
##
## Numbers of non-zeros for X super:
## comp1 comp2
##    2     2
##
## Numbers of non-zeros for Y:
##        comp1 comp2
## block1    2     2
## block2    2     3
##
```

```
## Numbers of non-zeros for Y super:
## comp1 comp2
##     2     2
```

The results of the first component are as follows.

```
par(mfrow=c(2,2), oma = c(0, 0, 2, 0))
plot(fit222, axes = 1, plottype="bar",
block="super", XY="X")
plot(fit222, axes = 2, plottype="bar",
block="super", XY="X")
plot(fit222, axes = 1, plottype="bar",
block="block", XY="X")
plot(fit222, axes = 2, plottype="bar",
block="block", XY="X")
```

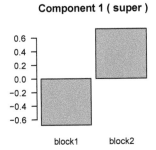

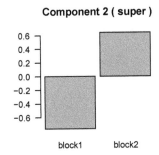

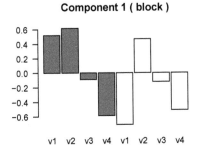

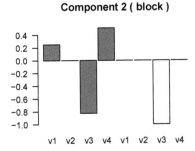

The results of the first component are as follows.

```
par(mfrow=c(2,2), oma = c(0, 0, 2, 0))
plot(fit222, axes = 1, plottype="bar",
block="super", XY="Y")
plot(fit222, axes = 2, plottype="bar",
block="super", XY="Y")
```

```
plot(fit222, axes = 1, plottype="bar",
block="block", XY="Y")
plot(fit222, axes = 2, plottype="bar",
block="block", XY="Y")
```

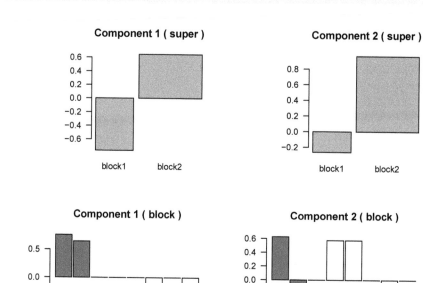

```
par(mfrow=c(1,2))
for(i in 1:2) plot(fit222, axes = i, XY="XY")
```

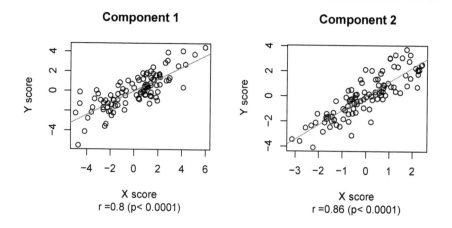

The relationship between the super score and the binary outcome Z is examined to compare the presence of supervision. The result with supervision is as follows.

```
par(mfrow=c(2,2))
for(xy in c("X","Y")){for(i in 1:2){
ss = fit222[[paste0("ss", xy)]][,i]
t1=t.test(ss~Z)
boxplot(ss~Z, main=paste(xy, "Comp", i),
sub=paste("t-test p =", round(t1$p.value,4)))
}}
```

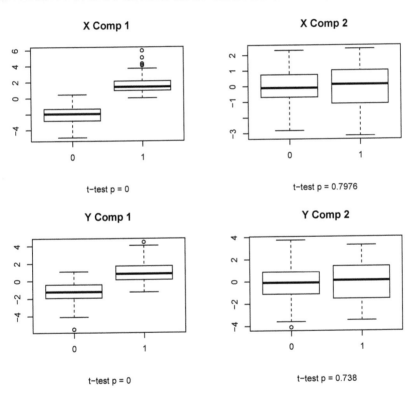

Independent component analysis for fMRI

Independent component analysis (ICA) is widely used for brain research as a very powerful and flexible data-driven analytical method. For an introduction to this method, please refer to Hyvärinen et al. (2001). Widely used general linear models require parameterization of data (e.g., brain reaction to stimulus), but ICA can be preformed only assuming independence. The interpretation of the results is thus left to the analyst. Image separation is an application scene in fMRI data analysis. Images showing brain activation sites observed by fMRI generally include artifacts caused by various signal sources, such as heartbeat and respiration, in addition to those related to the planned task; these artifacts hinder the clinical interpretation of the results.

ICA is widely used as a processing technique for this purpose. In fMRI data analysis, ICA can be roughly divided into two analyses. ICA assumes independence for data, but depending on whether it is spatial or temporal, it is divided into spatial ICA or temporal ICA. As it is considered easier to interpret spatially independent data, spatial ICA is widely used. Stone et al. (2002) proposes a spatial-temporal ICA.

An explanation of spatial ICA follows. The whole picture is shown in Figure 5.2. Let $T \times V$ data matrix of a subject be $\mathbf{X}$. Each row of this matrix corresponds to the voxel time series, V voxels, and T time points. $\mathbf{Y} \approx \mathbf{UDX}$, where $\mathbf{U}$ is the $T \times M$ orthogonal matrix and $\mathbf{D} = \mathrm{diag}(d_1, d_2, \ldots, d_M)$, $d_1 \geq d_2 \geq \cdots \geq d_M$, and $\mathbf{X}$ is the $M \times V$ orthogonal matrix. Suppose that the matrix $\mathbf{X}$ is decomposed into $\mathbf{X} = \mathbf{AS}$. Where $\mathbf{A}$ is a mixed matrix of $T \times M$ and $\mathbf{S}$ is the source matrix of $M \times V$. Each row of the source matrix is assumed to be statistically independent. In this decomposition, the k th column of $\mathbf{X}$ is equal to $\mathbf{A}$ multiplied by the k th column of $\mathbf{S}$. However, $k = 1, 2, \ldots, V$. In other words, each column of $\mathbf{X}$ is a mixture of M independent sources (independent components). In the independent component analysis, the objective is to find the mixing matrix $\mathbf{A}$. If the matrix $\mathbf{A}$ is obtained, it can be connected to the interpretation of the corresponding independent component from the waveform (period). In the example provided by Figure 5.2, since the third column of $\mathbf{A}$ has a cycle corresponding to the assignment, the region obtained by the independent component of $\mathbf{S}$ is the part that is activated by the task. In many cases, singular value decomposition is considered before ICA is applied, and dimension reduction is performed such as to make $\mathbf{X}$ rows number M. By doing this, we consider a square matrix of $M \times M$ as a mixing matrix.

There are many methods of estimating the mixing matrix, Calhoun and Adali (2006) shows the usefulness of Infomax (Bell and Sejnowski, 1995) in fMRI data analysis and becomes the default of the software GIFT. Kawaguchi et al. (2012) proposed a method based on the nonparametric density estimation and maximum likelihood method that showed superiority over existing methods such as Infomax through numerical experiments. The method based on the maximum likelihood method is performed as follows. $\mathbf{S}_k = (S_{1k}, S_{2k}, \ldots, S_{Mk})$, where $\mathbf{S}_k$ is the source vector in voxel k, $k = 1, 2, \ldots, V$. Each S_j has a density function f_j ($j = 1, 2, \ldots, M$). Then, the density function of $\mathbf{X}$ is $f_{\mathbf{X}}(\mathbf{x}) = \det(\mathbf{W}) \prod_{j=1}^{M} f_j(\mathbf{w}_j \mathbf{x})$. However, $\mathbf{W} = \mathbf{A}^{-1}$, called a reverse mixing matrix, $\mathbf{w}_j$ is the j th row of $\mathbf{W}$. Assume that the density function is specified by the parameter $\boldsymbol{\theta}$. The parameters of the inverse mixing matrix and the density function $(\mathbf{W}, \boldsymbol{\theta})$ are determined to

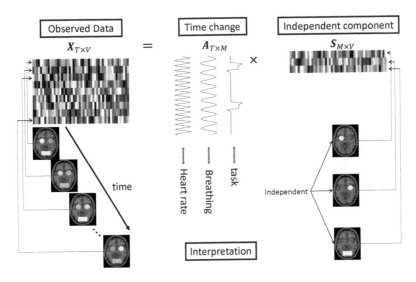

Figure 5.2: Spatial ICA for fMRI data.

maximize the following log-likelihood function.

$$\ell(\mathbf{W}, \boldsymbol{\theta}) = \sum_{k=1}^{V} \sum_{j=1}^{M} \log(f_j(\mathbf{w}_j^T \mathbf{x}_k)).$$

Please refer to Kawaguchi and Truong (2011) for an algorithm to maximize this function. As shown in Figure 5.3, since activation sites exist locally, the spatial distribution of fMRI data assumes a long tail distribution. Therefore, Kawaguchi et al. (2012) introduces a mixture density function that considers the features of fMRI data as a source density function. Kawaguchi and Truong (2016) provide a description of the overall method and its R-code.

For some unknown mixing ratio a,

$$f_j(x) = a f_{1j}(x) + (1-a) f_{2j}(x),$$

where the logarithm of $f_{1j}(x)$ is a polynomial spline density function and is expressed as follows.

$$\log(f_{1j}(x)) = C(\boldsymbol{\beta}_j) + \beta_{01j}x + \sum_{i=1}^{m_j} \beta_{1ij}(x - r_{ij})_+^3,$$

where $\boldsymbol{\beta}_j = (\beta_{01j}, \beta_{11j}, \ldots, \beta_{1m_jj})$ is the coefficient vector, $C(\boldsymbol{\beta}_j)$ is a normalization factor, r_{ij} is a knot and $(\mathbf{z})_+ = \max(\mathbf{z}, \mathbf{0})$. Further, $f_{2j}(x) = sech^2(x)/2$ is the logistic density function. In this model, f_{1j} represents the activation site

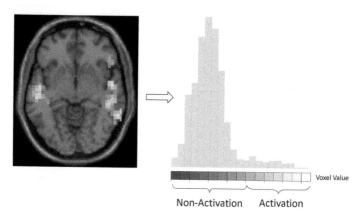

Figure 5.3: Distribution of activated and non-activated regions.

and f_{2j} represents the non-activation site. A flexible nonparametric density function method is applied to the activation site, and parametric density A function is applied.

In addition, an ICA algorithm for fMRI data has been developed. Lee et al. (2011) considers the correlation within the source in the framework of temporal ICA, link ICA of Groves et al. (2011) considers covariates. Although the above ICA is for one subject, a method called group ICA has been proposed to allow for guesswork among subject groups. Calhoun et al. (2001) proposed a singular value decomposition for each subject and one data matrix by stacking the obtained matrices and applying an ICA algorithm to the data matrix. This method can be executed with the software GIFT. Beckmann and Smith (2005) proposed a tensor probabilistic ICA using a tensor product consisting of time, space, and three directions of a subject as a mixing matrix. A comprehensive model based on a model $\mathbf{X} = \mathbf{AS} + \mathbf{E}$ that includes error term $\mathbf{E}$ has been proposed by Guo and Pagnoni (2008), which has been developed in Guo (2011) to accommodate various grouping methods. ICA is also important for cleaning (denoising) the fMRI signal (Caballero-Gaudes and Reynolds, 2017).

Similar to ICA, principal component analysis (PCA) is also applied in connection with ICA; however, in the analysis of fMRI, PCA is often applied in the framework of functional data analysis. By treating the data as a function of time and performing smoothing, variance is suppressed and estimation accuracy is improved (Tian, 2010). This so-called functional principal component analysis is used in Lange et al. (1999) and Viviani et al. (2005); it detects the principal components related to the problem column with high accuracy. Zipunnikov et al. (2011) applies functional principal components to segmentation. Aston and Kirch (2012) performs change point analysis using functional data analysis.

Network analysis

The brain forms a network, the parts of which have separate functions and communicate with one another. fMRI allows for the clarification of the mechanism underlying this functional connectivity and of neuropathologies. Network analysis of fMRI data is roughly divided into (1) functional connectivity (Keller et al., 2011) and (2) effective connectivity (Stephan and Friston, 2010). Both represent spatially arranged relationships among brain regions; however, functional connectivity, temporal correlation and effective relationships are represented by undirected and directed graphs, respectively.

fMRI data obtained during the resting state has garnered increasing attention. By revealing networks during the resting state, activation induced by a given task (Zhang and Raichle, 2010) can be compared to this network for both individual afflicted with diseases or their healthy counterparts (default mode network; DMN).The typical analysis flow is as shown in Figure 5.4.

The analyst typically calculates the average time series in the region of interest (ROI) consisting of multiple voxels based on anatomical knowledge and estimates the correlation coefficient between them. In functional relation analysis, a threshold is set for the correlation coefficient to create an adjacency matrix, which informs a graph (Bullmore et al., 1996). An alternative method uses correlation in the frequency domain, not the correlation coefficient (Sun et al., 2004).

In effective correlation analysis, methods based on models such as structural equation modeling (SEM) (Penny et al., 2004; Taniwaki et al., 2007; James et al.,

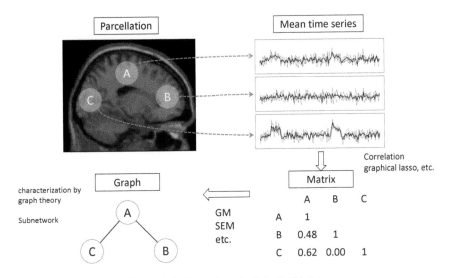

Figure 5.4: Network analysis in fMRI data.

2009) have been used in many cases. The Bayesian network (BN), which is employed in research on gene networks, is also useful in such studies (Zheng and Rajapakse, 2006); Rajapakse and Zhou (2007); Li et al. (2008) use the dynamic Bayesian network (Dynamic BN).

In recent years, methods combining SEM or BN with ICA are frequently used. The role of ICA in these studies is to substitute ROI. That is, the independent component image obtained by spatial ICA is divided into several areas, which are regarded as having some relevance to one another, and relations between these areas are drawn. Rajapakse et al. (2006) combines ICA with SEM, while Li et al. (2011) and Wu et al. (2011) combine BN with ICA. In addition, several studies have investigated the relationship between brain areas (Granger causality analysis, Goebel et al., 2003; Dynamic Causal Modelling, Friston et al., 2011; Daunizeau et al., 2011). Brain-imaging analytical methods for network analysis are summarized in van den Heuvel and Pol (2010) and Smith et al. (2011).

To avoid estimation in high dimensional space and improve interpretability, standard procedures define an area consisting of a collection of voxels sharing a similar time course. This predefined region is usually called a parcel. The total number of parcels in SHEN parcellation is 268 and is 264 in Power. The Human Brainnetome Atlas (Fan et al., 2016) containing 210 cortical and 36 subcortical ROIs.

Correlation analysis

Quantify the relationship between regions. This numerical value is stored as a matrix in which each region is arranged in rows and columns and the network is drawn by expressing it in a graph. Statistical association matrices include the variance-covariance matrix, the correlation matrix, the precision matrix, and the partial correlation matrix.

Since the region corresponds to the variable, the target is multivariate data. Therefore, the variance-covariance matrix is used as a basic statistical method for the interrelationship between variables. For two variate data $(x_1, y_1), (x_2, y_2), \ldots, (x_n, y_n)$, the covariance is $S_{xy} = (\sum_{i=1}^{n}(x_i - \bar{x})(y_i - \bar{y}))/(n - 1)$. Since the subscripts and variables in S_{xy} correspond, the formulas for the variance and covariance for the trivariate according to that definition are given as well and the variance-covariance matrix is

$$\Sigma = \begin{pmatrix} S_{xx} & S_{xy} & S_{xz} \\ S_{xy} & S_{yy} & S_{yz} \\ S_{xz} & S_{yz} & S_{zz} \end{pmatrix}$$

The diagonal element of the variance-covariance matrix is the variance of each variable and the off-diagonal element is the covariance. This can be converted to a correlation matrix by normalizing the diagonal elements to one and the off-diagonal elements to be correlation coefficients in the following form.

$$R = \begin{pmatrix} r_{xx} & r_{xy} & r_{xz} \\ r_{xy} & r_{yy} & r_{yz} \\ r_{xz} & r_{yz} & r_{zz} \end{pmatrix}, \quad r_{xy} = \frac{S_{xy}}{\sqrt{S_{xx}}\sqrt{S_{yy}}}$$

This is a relationship for each region pair and the effect of regions other than the pair is not considered. There is a partial correlation coefficient that quantifies the relationship when considering the influence of other regions. It starts by finding the inverse of the variance-covariance matrix, called the precision matrix.

$$\Theta = \Sigma^{-1} = \begin{pmatrix} S^{xx} & S^{xy} & S^{xz} \\ S^{xy} & S^{yy} & S^{yz} \\ S^{xz} & S^{yz} & S^{zz} \end{pmatrix}$$

When the variance-covariance matrix is standardized for this precision matrix, a partial correlation matrix is obtained in which off-diagonal elements have partial correlation coefficients.

$$\Psi = \begin{pmatrix} r^{xx} & r^{xy} & r^{xz} \\ r^{xy} & r^{yy} & r^{yz} \\ r^{xz} & r^{yz} & r^{zz} \end{pmatrix}, \quad r^{xy} = \frac{S^{xy}}{\sqrt{S^{xx}}\sqrt{S^{yy}}}$$

This partial correlation coefficient is equivalent to performing multiple regression analysis with a certain region as the objective variable and the remaining regions as explanatory variables. In other words, we will see the correlation excluding the influence between other regions and thus the method of partial correlation is often used in brain image analysis.

R example

Load the following package as preparation.

```
library(PerformanceAnalytics)
library(ppcor)
library(qgraph)
library(glasso)
library(corrplot)
```

The number of subjects is set at 20, generating data for three variables: x1 generates random numbers from a normal distribution, and x2 and x3 are generated from a regression model with x1 as the explanatory variable. In this way, the true correlations would be between x1 and x2 and between x1 and x3, but there would also be pseudo-correlations in x2 and x3.

```
n = 20
set.seed(21)
x1 = rnorm(n, 0, 0.1)
x2 = 0.8*x1 + rnorm(n, 0, 0.1)
x3 = 0.4*x1 + rnorm(n, 0, 0.1)
X = data.frame(cbind(x1,x2,x3))
```

The scatter plot and the correlation between three variables are displayed by using the chart.Correlation function of the PerformanceAnalytics package.

```
chart.Correlation(X)
```

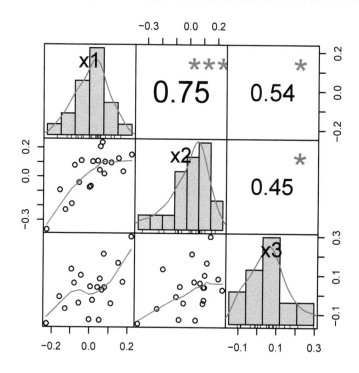

The variance-covariance matrix is computed by the var function.

```
(var1 = var(X))
```

```
##             x1          x2          x3
## x1 0.013018748 0.012951874 0.007004248
## x2 0.012951874 0.023073132 0.007874942
## x3 0.007004248 0.007874942 0.013153155
```

The correlation coefficient between X2 and X3 is obtained by standardizing the corresponding variance.

```
var1[2,3] / sqrt(var1[2,2]*var1[3,3])
```

```
## [1] 0.4520423
```

The middle strength of correlation was observed.

Although the above computation is available for the correlation matrix, it can be computed by the cor function.

```
(cor1 = cor(X))
```

```
##               x1         x2         x3
## x1 1.0000000 0.7472995 0.5352565
## x2 0.7472995 1.0000000 0.4520423
## x3 0.5352565 0.4520423 1.0000000
```

The element in the second row and the third column was the same as the result of the computation above.

The precision matrix is defined as the inverse of the correlation matrix and is computed as follows.

```
(prec1 = solve(cor1))
```

```
##               x1         x2         x3
## x1  2.547469 -1.6179573 -0.6321640
## x2 -1.617957  2.2844246 -0.1666345
## x3 -0.632164 -0.1666345  1.4136957
```

The partial correlation coefficient between X2 and X3 is obtained by standardizing the corresponding element of the precision matrix.

```
-prec1[2,3] / sqrt(prec1[2,2]*prec1[3,3])
```

```
## [1] 0.09272534
```

A weak correlation was observed against the correlation coefficient.

The partial correlation matrix is computed by the pcor function of the ppcor package.

```
(pcor1 = pcor(X, method="pearson")$estimate)
```

```
##               x1         x2         x3
## x1 1.0000000 0.67069368 0.33311742
## x2 0.6706937 1.00000000 0.09272534
## x3 0.3331174 0.09272534 1.00000000
```

The element in the second row and the third column was the same as the result of the computation above.

Multiple Regression

The observed difference between the correlation coefficient and the partial correlation coefficient can be explained by using the multiple regression model.

First, the single regression model with the response variable x2 and the explanatory variable x3 is applied.

```
lm1 = lm(x2~x3, data=X)
summary(lm1)
```

```
##
## Call:
## lm(formula = x2 ~ x3, data = X)
##
## Residuals:
##      Min       1Q   Median       3Q      Max
## -0.27625 -0.08688  0.01693  0.09547  0.22299
##
## Coefficients:
##               Estimate Std. Error t value Pr(>|t|)
## (Intercept) 5.527e-05  3.235e-02   0.002   0.9987
## x3          5.987e-01  2.785e-01   2.150   0.0454 *
## ---
## Signif. codes:
## 0 '***' 0.001 '**' 0.01 '*' 0.05 '.' 0.1 ' ' 1
##
## Residual standard error: 0.1392 on 18 degrees of freedom
## Multiple R-squared:  0.2043, Adjusted R-squared:  0.1601
## F-statistic: 4.623 on 1 and 18 DF,  p-value: 0.04539
```

The significant relationship at the 5% level was observed with p=0.0454.

On the other hand, the multiple regression model with the response variable x2 and the explanatory variables x3 and x1 is applied.

```
lm2 = lm(x2~x3+x1, data=X)
summary(lm2)
```

```
##
## Call:
## lm(formula = x2 ~ x3 + x1, data = X)
```

```
##
## Residuals:
##        Min          1Q     Median          3Q         Max
## -0.139583 -0.080430 -0.000494   0.068308   0.169906
##
## Coefficients:
##               Estimate Std. Error t value Pr(>|t|)
## (Intercept) 0.003106    0.024707   0.126  0.90143
## x3          0.096611    0.251610   0.384  0.70576
## x1          0.942885    0.252906   3.728  0.00167 **
## ---
## Signif. codes:
## 0 '***' 0.001 '**' 0.01 '*' 0.05 '.' 0.1 ' ' 1
##
## Residual standard error: 0.1062 on 17 degrees of freedom
## Multiple R-squared:  0.5623, Adjusted R-squared:   0.5108
## F-statistic: 10.92 on 2 and 17 DF,  p-value: 0.0008921
```

In this output, the non-significant relationship at the 5% level was observed with p=0.70576. Because the x2 and x3 were generated from the regression model with the explanatory variable x1 and the noise, the crude correlation between x2 and x3 would be observed, however, it would be reduced by adjusting x1, which is its cause. The partial correlation can also represent the relationship between variables adjusting the remaining variables.

Graphical lasso

Deciding whether to draw a line between areas or not is easier than in network construction. If penalized estimation is used, the cut can be determined by adjusting the penalty parameters. A typical example is the graphical lasso, which has recently been extended to Bayesian, dynamic and subgraph representations (Warnick et al., 2018). Here we introduce the basic graphical lasso method.

X_i is p-dimensional vector at i-th time following the multivariate normal distribution with zero mean vector and the $p \times p$ precision matrix Θ (the variance-covariance matrix Θ^{-1}) to be estimated, that is,

$$X_1,\ldots,X_n \sim N(0,\Theta^{-1}).$$

The density function for the multivariate normal distribution is given by

$$f(x|0,\Theta) = \frac{\det(\Theta)^{1/2}}{(2\pi)^{p/2}} \exp\left(-\frac{1}{2}x^\top \Theta x\right).$$

The log likelihood for Θ is given by

$$\log \prod_{i=1}^{n} f(x_i|0,\Theta) = \sum_{i=1}^{n} \left\{ -\frac{p}{2}\log(2\pi) + \frac{1}{2}\log(\det\Theta) - \frac{1}{2}x_i^{\top}\Theta x_i \right\}$$

$$= (constant) + \frac{n}{2} \left\{ \log(\det\Theta) - \mathrm{tr}\left(\Theta\frac{1}{n}\sum_{i=1}^{n} xx^{\top} \right) \right\}$$

$$= (constant) + \frac{n}{2} \left\{ \log(\det\Theta) - \mathrm{tr}(S\Theta) \right\}.$$

The objective function to estimate Θ is given as follows.

$$L(\Theta) = \log\det\Theta - \mathrm{tr}(S\Theta) - \lambda\|\Theta\|_1$$

where S is the sample variance-covariance matrix. The derivative of L is given by

$$\frac{\partial L(\Theta)}{\partial \Theta} = \Theta^{-1} - S - \lambda\,\mathrm{sign}(\Theta).$$

Thus, the following equation is solved to obtain the optimal parameters.

$$\Theta^{-1} - S - \lambda\,\mathrm{sign}(\Theta) = 0. \tag{5.3}$$

For a convenient computation, each matrix is decomposed as follows.

$$\Theta = \begin{pmatrix} \Theta_0 & \boldsymbol{\theta}_1 \\ \boldsymbol{\theta}_1^{\top} & \theta_2 \end{pmatrix}, \quad \Sigma = \Theta^{-1} = \begin{pmatrix} \Sigma_0 & \boldsymbol{\sigma}_1 \\ \boldsymbol{\sigma}_1^{\top} & \sigma_2 \end{pmatrix}, \quad S = \begin{pmatrix} S_0 & \boldsymbol{s}_1 \\ \boldsymbol{s}_1^{\top} & s_2 \end{pmatrix} \tag{5.4}$$

where $\boldsymbol{\theta}_1$, $\boldsymbol{\sigma}_1$, and $\boldsymbol{s}_1$ are $p-1$ dimensional vectors and θ_2, σ_2, and s_2 are scalars. From this, the algorithm to estimate Θ will fix Σ_0 to estimate $\boldsymbol{\theta}_1$ and θ_2. $\boldsymbol{\theta}_1$ and θ_2 are obtained by sequentially choosing from the original Θ column and rearranging them into the last column. In fact, $\boldsymbol{\theta}_1$ and θ_2 will be estimated from $\boldsymbol{\sigma}_1$ and σ_2, as can be seen below.

Using the decomposition (5.4), from the equation (5.3),

$$\boldsymbol{\sigma}_1 - \boldsymbol{s}_1 - \lambda\,\mathrm{sign}(\boldsymbol{\theta}_1) \tag{5.5}$$

and $\sigma_2 - s_2 - \lambda\,\mathrm{sign}(\theta_2)$ which yields $\sigma_2 = s_2 + \lambda$ because $\mathrm{sign}(\theta_2) = 1$.

In addition, the product $\Theta^{-1}\Theta = I$ is given as follows using the decomposition (5.4).

$$\Theta^{-1}\Theta = \begin{pmatrix} \Sigma_0 & \boldsymbol{\sigma}_1 \\ \boldsymbol{\sigma}_1^{\top} & \sigma_2 \end{pmatrix} \begin{pmatrix} \Theta_0 & \boldsymbol{\theta}_1 \\ \boldsymbol{\theta}_1^{\top} & \theta_2 \end{pmatrix} = \begin{pmatrix} 1 & 0 \\ 0 & 1 \end{pmatrix}. \tag{5.6}$$

From the upper right of (5.6), $\Sigma_0\boldsymbol{\theta}_1 + \boldsymbol{\sigma}_1\theta_2 = 0$ and yields

$$\boldsymbol{\theta}_1 = -\theta_2\Sigma_0^{-1}\boldsymbol{\sigma}_1. \tag{5.7}$$

From this equation,

$$\text{sign}(\boldsymbol{\theta}_1) = \text{sign}(-\theta_2 \Sigma_0^{-1} \boldsymbol{\sigma}_1) = -\text{sign}(\Sigma_0^{-1} \boldsymbol{\sigma}_1).$$

Substituting this into the equation (5.5),

$$\boldsymbol{\sigma}_1 - \boldsymbol{s}_1 + \lambda \,\text{sign}(\Sigma_0^{-1} \boldsymbol{\sigma}_1).$$

Putting $\Sigma_0^{-1} \boldsymbol{\sigma}_1 = \boldsymbol{\beta}$, we obtain $\boldsymbol{\sigma}_1 = \Sigma_0 \boldsymbol{\beta}$ and the equation (5.5) is

$$\Sigma_0 \boldsymbol{\beta} - \boldsymbol{s}_1 + \lambda \,\text{sign}(\boldsymbol{\beta}).$$

This is represented for the j-th element

$$\sum_{k \neq j} \sigma_{0jk} \beta_k + \sigma_{0jj} \beta_j - s_{1j} + \lambda \,\text{sign}(\beta_j) + \lambda \,\text{sign}(\beta_j) = 0.$$

Putting $Y_j = \sum_{k \neq j} \sigma_{0jk} \beta_k - s_{1j}$ and the expression is transformed

$$\sigma_{0jj} \beta_j + \lambda \,\text{sign}(\beta_j) = Y_j \Longrightarrow \beta_j = \frac{1}{\sigma_{0jj}} \text{sign}(Y_j)(|Y_j| - \lambda)_+.$$

Thus, $\boldsymbol{\sigma}_1 = \Sigma_0 \boldsymbol{\beta}$ where $\boldsymbol{\beta}$ has the j-th element β_j.

Substituting the equation (5.7) into the right lower part of the equation (5.6),

$$-\boldsymbol{\sigma}_1^{\top} \Sigma_0^{-1} \boldsymbol{\sigma}_1 \theta_2 + \sigma_2 \theta_2 = 1.$$

From this,

$$\theta_2 = \frac{1}{\sigma_2 - \boldsymbol{\sigma}_1^{\top} \Sigma_0^{-1} \boldsymbol{\sigma}_1}.$$

Thus, θ_2 is determined from $\boldsymbol{\sigma}_1$ and σ_2, and then $\boldsymbol{\theta}_1$ is determined from (5.7). The equations to calculate $\boldsymbol{\theta}_1$ and θ_2 were derived.

As mentioned earlier, the sample variance covariance matrix is set as the initial value of Σ, the decomposition is carried out by selecting one of the p columns and $\boldsymbol{\theta}_1$ and θ_2 are computed with the above equations repeating for p columns. This is further repeated until Θ and Σ converge. The resulting Θ and Σ are the final estimates. See Friedman et al. (2008) for more details.

R example

Correlation analysis is used in the network representation. In this case, the null correlation means that the network between variables does not exist and vice versa. On the other hand, the regularization technique, which includes the constrained parameter estimation is available. The graphical lasso algorithm has a theoretically justified solution and it depends on the regularization parameter. The glasso function of the glasso package implements the algorithm and the different parameters produce the following constrained partial correlation matrices.

```
rhos = c(0, 0.5, 0.7, 0.9)
par(mfrow=c(1,4))
for(rho1 in rhos){
MO = glasso(cor1, rho=rho1)$wi
M = -cov2cor(MO); diag(M)=1;
rownames(M)=colnames(M)=colnames(X)
corrplot.mixed(M, lower.col = "black", number.cex = 1,
title=paste("rho =", rho1),mar=c(0,0,1,0) )
l
```

The larger parameter (rho) has more zero correlations.

The optimized parameter is decided by extended Bayesian information criterion (EBIC), which can be implemented by the EBICglasso function of the qgraph package. Gamma is the tuning parameter of the EBIC and 0.5 is recommended; setting it to 0 will use the conventional BIC.

```
CorMat = cor_auto(as.matrix(X))
(EBICgraph = EBICglasso(CorMat, nrow(X), gamma=0.5,
threshold = TRUE))
```

```
## Note: Network with lowest lambda selected as best network:
   assumption of sparsity might be violated.
```

```
##              x1          x2          x3
## x1 0.0000000 0.6640005 0.3299881
## x2 0.6640005 0.0000000 0.0000000
## x3 0.3299881 0.0000000 0.0000000
```

The partial correlation between x2 and x3 was constrained in the optimized result.

The output was represented as a network as follows.

```
EBICgraph2 = qgraph(EBICgraph, layout = "spring", title = "")
plot(EBICgraph2)
```

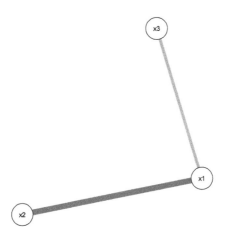

The edges represent not only the existance but also the strength of the relationship between variables.

Graph Theory

Once the graph is drawn, the graph structure can be examined later. Many pioneering studies have applied graph theory as a complex network analysis (Rubinov and Sporns, 2010, Guye et al., 2010, Bullmore and Sporns, 2009, Lo et al., 2011). In complex network analysis, it is characterized by feature values such as cluster coefficient and average distance calculated from the graph.

The middle of the lattice and random graphs is called "small world network" (Watts and Strogatz, 1998), and the brain network shares this small world property. It is characterized by high cluster coefficients and short mean distances. Sanz-Arigita et al. (2010) reported that the brain networks of patients with Alzheimer's disease collapses small-world properties and thus resembles a random graph.

Meta-analysis

Meta-analysis is a statistical method for obtaining strong evidence by integrating the results of published research. In this section, we introduced meta-analysis specialized for brain image analysis. Although brain image analysis plays an important role in neurological disease research, it is difficult to acquire adequate sample sizes to obtain reliable statistical analysis results because images are generally expensive. It is thought that by using meta-analysis, research results from a small number of samples could be combined to obtain more reliable results. In this book, we introduced the method and software to execute the coordinate based meta-analysis based on image coordinates and that significant difference is recognized especially in brain image meta-analysis.

Overview

In preparation for meta-analysis, there is a systematic review to collect documents without bias and the effect size that is the result of statistical analysis is taken from that document (for example, if the evaluation item is a continuous amount, the difference between groups in the average value). In meta-analysis, effect sizes in individual studies are integrated and new analysis results are obtained. Refer to Schwarzer et al. (2015) for general meta-analysis.

Brain images are divided into three-dimensional spaces in units called voxels (pixels in the case of 2D) and analyzed based on the signals (voxel values) recorded for each voxel. For example, in the case of a structured brain image, focusing on a certain area, the volume is calculated from the voxel value of that area and the form such as atrophy is evaluated. Another analysis (Voxel-based morphometry, VBM) compares all voxel values between the diseased group and the healthy group for the entire brain.

If the evaluation index is a brain volume value calculated from an image, it is a continuous volume, so general meta-analysis can be applied as described above. As a result, VBM can obtain a statistical image (statistical map) consisting of effect quantities for each voxel of the whole brain and perform meta-analysis and these are classified as Image-based meta-analysis. However, to obtain statistical images from each study, it is necessary to contact individual researchers, which is difficult and it is also easier to collect information described in the published paper. Therefore, there are many studies in which the coordinates on the standard brain (referred to as foci) representing the anatomical position of the brain where significant effects (such as differences between groups) were recognized in each study are the targets of integration in brain image analysis. Such a method is called coordinate based meta-analyses (CBMA) and is described in detail in the next section. For a review of meta-analysis of brain images (including general ones), see Radua and Mataix-Cols (2012).

As an application example of meta-analysis in neurological diseases, brain morphology of Alzheimer's disease or semantic dementia was compared with that of healthy Chapleau et al. (2016) and atrophy of the visual cortex was evaluated for

Alzheimer's disease and mild cognitive impairment Deng et al. (2016). A meta-analysis of gray matter atrophy has been conducted for subtypes of Parkinson's disease Yu et al. (2015). Both apply CBMA for VBM.

Coordinate based meta-analysis (CBMA)

CBMA is a method for integrating coordinates on a standard brain (Montreal Neurological Institute (MNI) template, Talairach) that has been reported from multiple studies. A typical example is the Activation Likelihood Estimate (ALE) (Eickhoff et al., 2009; Eickhoff et al., 2012; Turkeltaub et al., 2002; Turkeltaub et al., 2012). A similar approach is Kernel Density Analysis (KDA) (Wager et al., 2004). These are summarized in reviews Salimi-Khorshidi et al. (2009) and Kober and Wager (2010). There is also the evolution of a method termed signed differential mapping (SDM) (Radua and Mataix-Cols, 2009; Radua et al., 2010). The following introduces the related methods while introducing mainly ALE.

ALE is a method developed by Turkeltaub et al. (2002), but several improvements have been made since then. The results (the difference between groups) in image analysis appear as a cluster consisting of multiple voxels in the brain space. The coordinates reported in each study are the peaks in the cluster and represent one voxel in the cluster. Thus, when performing meta-analysis, clusters are reproduced in space based on coordinates. Suppose that several coordinates are reported within a study. Apply a Gaussian function (normal density function) centered at that coordinate. This represents a spatial probability map and is called a "modeled activation" (MA) map. To construct this map, Gaussian functions were simply added together in the initial ALE, but if very close coordinates were reported in one study, a high probability map was generated as a result when they were added together. To correct this, an integration method is proposed in which the maximum function value is used for coordinates where multiple Gaussian functions overlap and this is supported in the latest version of the software described later Turkeltaub et al. (2012). A map obtained by integrating probability maps created for each coordinate between studies is called an ALE map. Thus, ALE integrates probability maps reconstructed from coordinates rather than integrating coordinates. KDA applies a spherical kernel function (one of which is a Gaussian function), but it shows whether there are some peaks in the radius. In SDM, coordinates representing positive and negative effects (gray matter increase and decrease) are simultaneously reconstructed into the same map. In other parts, ALE, KDA, and SDM are considered to be almost the same.

In general, in meta-analysis, the simplest method for integrating research results is to average the effect sizes, but it is possible to obtain results that are more accurate by focusing on quality research. There is a weighted average method for this purpose. The Gaussian function used in ALE is specified by a center representing position and a full width at half maximum (FWHM) representing width. When creating an MA map, the center is the reported coordinate. By making FWHM inversely proportional to the sample size in the study, ALE creates MA maps that shows a high probability with a narrow range around the coordinates reported from studies with large sample sizes (Eickhoff et al., 2009).

Figure 5.5 illustrates ALE as an example. Assume that 2, 2 and 3 coordinates are obtained from three studies, respectively, as shown in Figure 5.5 (1). As shown in Figure 5.5 (2), because the sample size in Study 1 is the largest (N = 100), a Gaussian function with a corresponding FWHM is applied and it can be seen that narrower and higher (which represents neighboring and high probability) Gaussian function than in Study 2 and Study 3 is applied. In Figure 5.5 (3), the MA map is obtained by taking the maximum value of the Gauss function in each study for each coordinate. In Figure 5.5 (4), the MA maps obtained in the three studies are now added together for each coordinate and an ALE map is obtained.

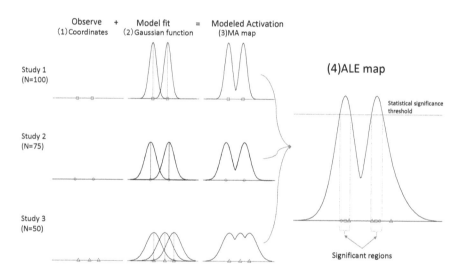

Figure 5.5: Schematic explanation of ALE.

Because brain imaging analysis is often a small sample or exploratory study, some of the reported coordinates may include chance. Since the ALE map is considered the same as the statistical image obtained by VBM, it is necessary to set a threshold for expressing statistical significance. A Permutation test is usually used for this purpose. At this time, the null hypothesis is that the peak coordinates are uniformly distributed throughout the brain, so calculate the ALE map after arranging the coordinates randomly and repeat this (for example, 10,000 times). This creates a null distribution for each voxel. If the original ALE map is located at the tail of this null distribution, it is determined that there is a significant difference. In this case, for example, if the tail is set to the upper 5% point or higher, the threshold value becomes the upper 5% point and the significance level is 5%. Eickhoff et al. (2012) discussed multiple corrections such as cluster estimation as a method of determining the threshold. For example, suppose that the threshold shown in Figure 5.5 (4) is obtained. A region corresponding to an ALE map higher than this threshold is a significant region. Here, it can be seen that coordinates with dense research or research with a large sample size have a high probability of ALE maps and tend to be above the threshold.

In some cases, the true value of the difference between groups varies in each study. Possible causes include potential deviations such as differences in research design and differences in facilities. It is necessary to adapt the analysis method depending on whether there is a deviation in the true value. In general meta-analysis, there are test methods and scales to confirm homogeneity between the studies. If heterogeneity is found between the studies, analyses using a random-effects model is performed. There is a DerSimonian-Laird method. In CBMA, the same can be said for the reported coordinates when there are uncertainties because the coordinates are related to the resolution. In ALE, an improvement that adopts FWHM considering such variations has been proposed Eickhoff et al. (2009), which corresponds to the random-effects model in ALE. Multi-level Kernel Density Analysis (MKDA) was proposed as a random effects model in KDA Wager et al. (2007).

Software and database

Software for executing the method introduced in the previous section has been released. For ALE, it can be executed by GingerALE software Version 3.0.2 (`https://brainmap.org/ale/`). Figure 5.6 provides an overview.

First, prepare an input file for GingerALE in the format shown in Figure 5.6 (1). The comment after the double slash is the standard brain on the first line, the research name, the research year, the name of the comparison on the second line and the sample size on the third line. Write the coordinates on the line below. Each study must be distinguished by a line break. This file can be executed by loading it with GingerALE like Figure 5.6 (2) and pressing the Compute button.

(1) input text file (ALE prepare)   (2) GingerALE Software (ALE implementation)

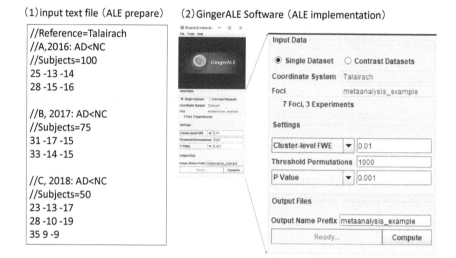

Figure 5.6: ALE implementation by the software.

The ALE map, which is the output of GingerALE, is overlaid on the template.

```
x0 = readNIfTI("metaanalysis_example_ALE.nii",
reorient = FALSE)
x = sizechange(x0, refsize=dim(template))
coat(template, x)
```

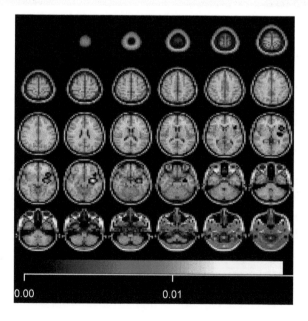

Similarly, an ALE map above the threshold, which is the output of GingerALE, is overlaid on the template.

```
x0 = readNIfTI("metaanalysis_example_C01_1k_ALE.nii",
 reorient = FALSE)
x = sizechange(x0, refsize=dim(template))
coat(template, x)
```

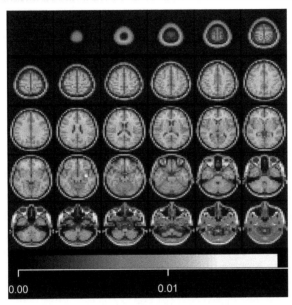

The names of the detected regions are presented as follows.

```
atlastable(tmpatlas, x, atlasdataset)
```

##	ROIid	ROIname	sizepct
## 37	37	Left Hippocampus	0.108
## 0	0	Background	0.000
## 73	73	Left Putamen	0.067
## 41	41	Left Amygdala	0.222
## 81	81	Left Superior Temporal	0.010
## 1	1	Left Precentral	0.000
## 2	2	Right Precentral	0.000
## 3	3	Left Superior Frontal	0.000
## 4	4	Right Superior Frontal	0.000
## 5	5	Left Superior Frontal Orbital	0.000

##	sumvalue	Min.	Mean	Max.
## 37	0.042	0	0	0.017
## 0	0.083	0	0	0.016

```
## 73    0.020    0    0 0.011
## 41    0.018    0    0 0.011
## 81    0.007    0    0 0.007
## 1     0.000    0    0 0.000
## 2     0.000    0    0 0.000
## 3     0.000    0    0 0.000
## 4     0.000    0    0 0.000
## 5     0.000    0    0 0.000
```

For other methods, MKDA or KDA can be run by MKDA and KDA toolboxes Meta-analysis Toolbox (http://www.columbia.edu/cu/psychology/tor/meta-analysis.html). Furthermore, IBMA: An SPM toolbox for Neuroimaging Image-Based Meta-Analysis (https://github.com/NeuroimagingMetaAnalysis/ibma) is also available for Image-based meta-analysis introduction.

Each research result, which is the original data of meta-analysis, is accumulated as a database. Peak coordinates are collected in BrainMap (www.brainmap.org), NeuroSynth (www.neurosynth.org). There are many others and they are summarized in Radua and Mataix-Cols (2012), so please refer to the details. Recently, coordinates and statistical images have been accumulated for various modalities, neurovault (http://neurovault.org/, Gorgolewski et al., 2016) and the reported areas have been accumulated for use in seed-based connectivity analysis. Archive of Neuroimaging Meta-analyses (ANIMA) (http://anima.fz-juelich.de/, Reid et al., 2016), the Brain Analysis Library of Spatial maps and Atlases (BALSA) (https://balsa.wustl.edu/, Van Essen et al., 2017).

Summary

A meta-analysis that integrates the results of brain image analysis was introduced. The method based on coordinates in the standard brain is the mainstream because it is easy to collect data and the method is being established and software that can be executed has been released. There are several analyses of developments. One is meta-analysis for multi-modal analysis that analyzes various images in an integrated manner Bora et al. (2011) Radua et al. (2013) Liu et al. (2015). Another example is network analysis, but a method called co-activation patterns has been developed by Laird et al. (2013) and Xue et al. (2014), Caspers et al. (2014) also developed PaMiNI software (http://www.fz-juelich.de/inm/inm-1/DE/Service/Download/download_node.html), which is software for executing as well as developing methods. Additionally, a brain image meta-analysis using a statistical model based on a hierarchical Bayesian cluster process has been developed as an approach different from the method based on the kernel functions such as the Gaussian function introduced in this section (Kang et al., 2011; Kang et al., 2014; Montagna et al., 2018). The article is summarized by Samartsidis et al. (2017).

Additionally, when writing a paper, generally only good results tend to be published. Because meta-analysis only uses published research, there may be publication bias in the results. In general meta-analysis, there are methods of evaluation (Funnel Plot) and correction (Trim-fill method). When a systematic review is required, the results of sensitivity analysis should be shown as evidence. Sensitivity analysis is a study of how much the meta-analysis results change depending on whether or not a particular study is included. These methods are also necessary for brain image meta-analysis.

In summary, the brain image meta-analysis can collect the results of each study obtained from small samples for stronger evidence. This is expected to be useful for the verification of clinical hypotheses because the database is being developed and the software executables are substantial and clinically valuable.

References

Aggarwal, C. C. and Reddy, C. K. (2014). Data clustering. *Algorithms and applications. Chapman & Hall/CRC Data mining and Knowledge Discovery series, Londra.*

Allen, G. I., Grosenick, L. and Taylor, J. (2014). A generalized least-square matrix decomposition. *Journal of the American Statistical Association* 109: 145–159.

Araki, Y. and Kawaguchi, A. (2019). Functional logistic discrimination with sparse pca and its application to the structural mri. *Behaviormetrika* 46: 147–162.

Araki, Y., Kawaguchi, A. and Yamashita, F. (2013). Regularized logistic discrimination with basis expansions for the early detection of alzheimer's disease based on three-dimensional mri data. *Advances in Data Analysis and Classification* 7: 109–119.

Arbabshirani, M. R., Plis, S., Sui, J. and Calhoun, V. D. (2017). Single subject prediction of brain disorders in neuroimaging: promises and pitfalls. *Neuroimage* 145: 137–165.

Ashburner, J. (2007). A fast diffeomorphic image registration algorithm. *NeuroImage* 38: 95–113.

Ashburner, J. (2012). Spm: a history. *Neuroimage* 62: 791–800.

Ashburner, J. and Friston, K. J. (2000). Voxel-based morphometry-the methods. *NeuroImage* 11: 805–821.

Ashburner, J. and Friston, K. J. (2005a). Unified segmentation. *Neuroimage* 26: 839–851.

Ashburner, J. and Friston, K. J. (2005b). Unified segmentation. *NeuroImage* 26: 839–851.

Ashby, F. G. (2019). *Statistical Analysis of fMRI Data*. MIT Press.

Aston, J. A. D. and Kirch, C. (2012). Evaluating stationarity via change-point alternatives with applications to fMRI data. *Annals of Applied Statistics* In Press.

Avants, B. B., Epstein, C. L., Grossman, M. and Gee, J. C. (2008). Symmetric diffeomorphic image registration with cross-correlation: Evaluating automated labeling of elderly and neurodegenerative brain. *Medical Image Analysis* 12: 26–41.

Avants, B. B., Tustison, N. and Song, G. (2009). Advanced normalization tools (ants). *Insight J* 2: 1–35.

Avants, B. B., Tustison, N. J., Song, G., Cook, P. A., Klein, A. and Gee, J. C. (2011). A reproducible evaluation of ANTs similarity metric performance in brain image registration. *Neuroimage* 54: 2033–2044.

Basaia, S., Agosta, F., Wagner, L., Canu, E., Magnani, G., Santangelo, R., Filippi, M., Initiative, A. D. N. et al. (2019). Automated classification of alzheimer's disease and mild cognitive impairment using a single mri and deep neural networks. *NeuroImage: Clinical* 21: 101645.

Beckmann, C. F. and Smith, S. M. (2005). Tensorial extensions of independent component analysis for multisubject FMRI analysis. *Neuroimage* 25: 294–311.

Bell, A. J. and Sejnowski, T. J. (1995). An information maximisation approach to blind separation and blind deconvolution. *Neural Computation* 7: 1129–1159.

Benjamini, Y. and Hochberg, Y. (1995). Controlling the false discovery rate: A practical and powerful approach to multiple testing. *Journal of the Royal Statistical Society. Series B (Methodological)* 57: 289–300.

Bookstein, F. (1989). Principal warps: Thin-plate splines and the decomposition of deformations. *IEEE Transactions on Pattern Analysis and Machine Intelligence* 11: 567–585.

Bora, E., Fornito, A., Radua, J., Walterfang, M., Seal, M., Wood, S. J., Yücel, M., Velakoulis, D. and Pantelis, C. (2011). Neuroanatomical abnormalities in schizophrenia: A multimodal voxelwise meta-analysis and meta-regression analysis. *Schizophrenia Research* 127: 46–57.

Buckner, R. L., Bandettini, P. A., O'Craven, K. M., Savoy, R. L., Petersen, S. E., Raichle, M. E. and Rosen, B. R. (1996). Detection of cortical activation during averaged single trials of a cognitive task using functional magnetic resonance imaging. *Proceedings of the National Academy of Sciences of the U.S.A.* 93: 14878–14883.

Bullmore, E. and Sporns, O. (2009). Complex brain networks: graph theoretical analysis of structural and functional systems. *Nature Reviews Neuroscience* 10: 186–198.

Bullmore, E. T., Rabe-Hesketh, S., Morris, R. G., Williams, S. C., Gregory, L., Gray, J. A. and Brammer, M. J. (1996). Functional magnetic resonance image analysis of a large-scale neurocognitive network. *Neuroimage* 4: 16–33.

Caballero-Gaudes, C. and Reynolds, R. C. (2017). Methods for cleaning the bold fmri signal. *Neuroimage* 154: 128–149.

Cabezas, M., Oliver, A., Lladó, X., Freixenet, J. and Cuadra, M. B. (2011). A review of atlas-based segmentation for magnetic resonance brain images. *Computer Methods and Programs in Biomedicine* 104: e158–e177.

Cadima, J. and Jolliffe, I. T. (1995). Loading and correlations in the interpretation of principle components. *Journal of Applied Statistics* 22: 203–214.

Calhoun, V. D. and Adali, T. (2006). Unmixing fmri with independent component analysis. *Engineering in Medicine and Biology Magazine, IEEE* 25: 79–90.

Calhoun, V. D., Adali, T., Pearlson, G. D. and Pekar, J. J. (2001). A method for making group inferences from functional mri data using independent component analysis. *Human Brain Mapping* 14: 140–151.

Casanova, R., Hsu, F.-C., Espeland, M. A., Initiative, A. D. N. et al. (2012). Classification of structural mri images in alzheimer's disease from the perspective of ill-posed problems. *PloS One* 7.

Casanova, R., Wagner, B., Whitlow, C. T., Williamson, J. D., Shumaker, S. A., Maldjian, J. A. and Espeland, M. A. (2011). High dimensional classification of structural mri alzheimer's disease data based on large scale regularization. *Frontiers in Neuroinformatics* 5: 22.

Caspers, J., Zilles, K., Beierle, C., Rottschy, C. and Eickhoff, S. B. (2014). A novel meta-analytic approach: Mining frequent co-activation patterns in neuroimaging databases. *Neuroimage* 90: 390–402.

Chapleau, M., Aldebert, J., Montembeault, M. and Brambati, S. M. (2016). Atrophy in alzheimer's disease and semantic dementia: An ale meta-analysis of voxel-based morphometry studies. *Journal of Alzheimer's Disease* 54: 941–955.

Chen, J. E. and Glover, G. H. (2015). Functional magnetic resonance imaging methods. *Neuropsychology Review* 25: 289–313.

Chumbley, J., Worsley, K., Flandin, G. and Friston, K. (2010). Topological fdr for neuroimaging. *NeuroImage* 49: 3057–3064.

Cuingnet, R., Gerardin, E., Tessieras, J., Auzias, G., Lehóricy, S., Habert, M. O., Chupin, M., Benali, H. and Colliot, O. (2011). Automatic classification of patients with alzheimer's disease from structural mri: A comparison of ten methods using the adni database. *NeuroImage* 56: 766–781.

Daunizeau, J., David, O. and Stephan, K. E. (2011). Dynamic causal modelling: A critical review of the biophysical and statistical foundations. *Neuroimage* 58: 312–322.

Deng, Y., Shi, L., Lei, Y., Liang, P., Li, K., Chu, W. C., Wang, D., Initiative, A. D. N. et al. (2016). Mapping the "what" and "where" visual cortices and their atrophy in alzheimer's disease: Combined activation likelihood estimation with voxel-based morphometry. *Frontiers in Human Neuroscience* 10: 333.

Ding, C., He, X. and Simon, H. D. (2005). On the equivalence of nonnegative matrix factorization and spectral clustering. In *Proceedings of the 2005 SIAM International Conference on Data Mining*, pp. 606–610. SIAM.

Dmitrienko, A. and D'Agostino Sr, R. (2013). Traditional multiplicity adjustment methods in clinical trials. *Statistics in Medicine* 32: 5172–5218.

Dryden, I. L. and Mardia, K. (1998). *Statistical Shape Analysis*. Wiley Series in Probability and Statistics: Probability and Statistics. J. Wiley.

Ecker, C., Marquand, A., Mourao-Miranda, J., Johnston, P., Daly, E. M., Brammer, M. J., Maltezos, S., Murphy, C. M., Robertson, D., Williams, S. C. and Murphy, D. G. (2010). Describing the brain in autism in five dimensions–magnetic resonance imaging-assisted diagnosis of autism spectrum disorder using a multiparameter classification approach. *J. Neurosci.* 30: 10612–10623.

Eickhoff, S. B., Bzdok, D., Laird, A. R., Kurth, F. and Fox, P. T. (2012). Activation likelihood estimation meta-analysis revisited. *Neuroimage* 59: 2349–2361.

Eickhoff, S. B., Laird, A. R., Grefkes, C., Wang, L. E., Zilles, K. and Fox, P. T. (2009). Coordinate-based activation likelihood estimation meta-analysis of neuroimaging data: A random-effects approach based on empirical estimates of spatial uncertainty. *Human Brain Mapping* 30: 2907–2926.

Fan, J. and Li, R. (2001). Variable selection via nonconcave penalized likelihood and its oracle properties. *Journal of the American Statistical Association* 96: 1348–1360.

Fan, L., Li, H., Zhuo, J., Zhang, Y., Wang, J., Chen, L., Yang, Z., Chu, C., Xie, S., Laird, A. R. et al. (2016). The human brainnetome atlas: a new brain atlas based on connectional architecture. *Cerebral Cortex* 26: 3508–3526.

Fan, Y., Shen, D., Gur, R. C., Gur, R. E. and Davatzikos, C. (2007). Compare: Classification of morphological patterns using adaptive regional elements. *IEEE Trans. Med. Imaging* 26: 93–105.

Feng, D., Tierney, L. and Magnotta, V. (2012). Mri tissue classification using high-resolution bayesian hidden markov normal mixture models. *Journal of the American Statistical Association* 107: 102–119.

Filippi, M. (2009). *FMRI Techniques and Protocols*. Springer Protocols. New York: Humana Press.

Fischl, B. (2012). Freesurfer. *Neuroimage* 62: 774–781.

Friedman, J., Hastie, T. and Tibshirani, R. (2008). Sparse inverse covariance estimation with the graphical lasso. *Biostatistics* 9: 432–441.

Frisoni, G. B., Boccardi, M., Barkhof, F., Blennow, K., Cappa, S., Chiotis, K., Démonet, J.-F., Garibotto, V., Giannakopoulos, P., Gietl, A. et al. (2017). Strategic roadmap for an early diagnosis of alzheimer's disease based on biomarkers. *The Lancet Neurology* 16: 661–676.

Frisoni, G. B., Fox, N. C., Jack, C. R., Scheltens, P. and Thompson, P. M. (2010). The clinical use of structural mri in alzheimer disease. *Nature Reviews Neurology* 6: 67–77.

Friston, K., Ashburner, J., Kiebel, S., Nichols, T. and Penny, W. (2007). *Statistical Parametric Mapping: The Analysis of Functional Brain Images*. London: Academic Press.

Friston, K., Josephs, O., Rees, G. and Turner, R. (1998). Non-linear event-related responses in fMRI. *Magnetic Resonance in Medicine* 39: 41–52.

Friston, K. J., Li, B., Daunizeau, J. and Stephan, K. E. (2011). Network discovery with DCM. *Neuroimage* 56: 1202–1221.

Friston, K. J., Worsley, K. J., Frackowiak, R. S. J., Mazziotta, J. C. and Evans, A. C. (1994). Assessing the significance of focal activations using their spatial extent. *Human Brain Mapping* 1: 214–220.

Genovese, C., Lazar, N. and Nichols, T. (2002). Thresholding of statistical maps in functional neuroimaging using the false discovery rate. *NeuroImage*, pp. 870–878.

Goebel, R., Roebroeck, A., Kim, D. S. and Formisano, E. (2003). Investigating directed cortical interactions in time-resolved fMRI data using vector autoregressive modeling and Granger causality mapping. *Magn. Reson. Imaging* 21: 1251–1261.

Goeman, J. J. and Solari, A. (2014). Multiple hypothesis testing in genomics. *Statistics in Medicine* 33: 1946–1978.

Gorgolewski, K. J., Varoquaux, G., Rivera, G., Schwartz, Y., Sochat, V. V., Ghosh, S. S., Maumet, C., Nichols, T. E., Poline, J.-B., Yarkoni, T. et al. (2016). Neurovault. org: A repository for sharing unthresholded statistical maps, parcellations, and atlases of the human brain. *Neuroimage* 124: 1242–1244.

Gross, J. (2019). Magnetoencephalography in cognitive neuroscience: A primer. *Neuron* 104: 189–204.

Groves, A. R., Beckmann, C. F., Smith, S. M. and Woolrich, M. W. (2011). Linked independent component analysis for multimodal data fusion. *Neuroimage* 54: 2198–2217.

Guo, Y. (2011). A general probabilistic model for group independent component analysis and its estimation methods. *Biometrics* 67: 1532–1542.

Guo, Y. and Pagnoni, G. (2008). A unified framework for group independent component analysis for multi-subject fMRI data. *Neuroimage* 42: 1078–1093.

Guye, M., Bettus, G., Bartolomei, F. and Cozzone, P. (2010). Graph theoretical analysis of structural and functional connectivity mri in normal and pathological brain networks. *Magnetic Resonance Materials in Physics, Biology and Medicine* 23: 409–421.

Hackmack, K., Paul, F., Weygandt, M., Allefeld, C. and Haynes, J.-D. (2012). Multi-scale classification of disease using structural mri and wavelet transform. *NeuroImage* 62: 48–58.

Haxby, J. V., Connolly, A. C. and Guntupalli, J. S. (2014). Decoding neural representational spaces using multivariate pattern analysis. Annual Review of Neuroscience 37: 435–456.

Haynes, J., Sakai, K., Rees, G., Gilbert, S., Frith, C. and Passingham, R. (2007). Reading hidden intentions in the human brain. *Current Biology* 17: 323–328.

Heinsfeld, A. S., Franco, A. R., Craddock, R. C., Buchweitz, A. and Meneguzzi, F. (2018). Identification of autism spectrum disorder using deep learning and the abide dataset. *NeuroImage: Clinical* 17: 16–23.

Hyvärinen, A., Karhunen, J. and Oja, E. (2001). *Independent Component Analysis*. John Wiley & Sons.

James, G. A., Kelley, M. E., Craddock, R. C., Holtzheimer, P. E., Dunlop, B. W., Nemeroff, C. B., Mayberg, H. S. and Hu, X. P. (2009). Exploratory structural equation modeling of resting-state fmri: Applicability of group models to individual subjects. *NeuroImage* 45: 778–787.

Jenkinson, M., Beckmann, C. F., Behrens, T. E., Woolrich, M. W. and Smith, S. M. (2012). Fsl. *Neuroimage* 62: 782–790.

Jolliffe, I. T., Trendafilov, N. T. and Uddin, M. (2003). A modified principal component technique based on the lasso. *Journal of Computational and Graphical Statistics* 12: 531–547.

Joo, S. H., Lim, H. K. and Lee, C. U. (2016). Three large-scale functional brain networks from resting-state functional mri in subjects with different levels of cognitive impairment. *Psychiatry Investigation* 13: 1.

Jovicich, J., Czanner, S., Greve, D., Haley, E., van der Kouwe, A., Gollub, R., Kennedy, D., Schmitt, F., Brown, G., Macfall, J., Fischl, B., and Dale, A. (2006). Reliability in multi-site structural MRI studies: effects of gradient non-linearity correction on phantom and human data. *NeuroImage* 30: 436–443.

Kang, J., Johnson, T. D., Nichols, T. E. and Wager, T. D. (2011). Meta analysis of functional neuroimaging data via bayesian spatial point processes. *Journal of the American Statistical Association* 106: 124–134.

Kang, J., Nichols, T. E., Wager, T. D. and Johnson, T. D. (2014). A bayesian hierarchical spatial point process model for multi-type neuroimaging meta-analysis. *The Annals of Applied Statistics* 8: 1800.

Kawaguchi, A. (2019). Supervised sparse components analysis with application to brain imaging data. In *Neuroimaging*. IntechOpen.

Kawaguchi, A. and Truong, K. Y. (2011). Logspline independent component analysis. *Bulletin of Informatics and Cybernetics* 43: 83–94.

Kawaguchi, A. and Truong, Y. K. (2016). Polynomial spline independent component analysis with application to fmri data. pp. 229–264. *In*: Truong, Y. K. and Lewis, M. M. (eds.). *Statistical Techniques for Neuroscientists*. CRC Press.

Kawaguchi, A., Truong, Y. K. and Huang, X. (2012). Application of polynomial spline independent component analysis to fmri data. pp. 197–208. *In*: Naik, G. (ed.). *Independent Component Analysis for Audio and Biosignal Applications*. Intech.

Kawaguchi, A. and Yamashita, F. (2017). Supervised multiblock sparse multivariable analysis with application to multimodal brain imaging genetics. *Biostatistics* 18: 651–665.

Keller, C. J., Bickel, S., Entz, L., Ulbert, I., Milham, M. P., Kelly, C. and Mehta, A. D. (2011). Intrinsic functional architecture predicts electrically evoked responses in the human brain. *Proc. Natl. Acad. Sci. U.S.A.* 108: 10308–10313.

Kiebel, S. J., Poline, J. B., Friston, K. J., Holmes, A. P. and Worsley, K. J. (1999). Robust smoothness estimation in statistical parametric maps using standardized residuals from the general linear model. *Neuroimage* 10: 756–766.

Klein, A., Andersson, J., Ardekani, B. A., Ashburner, J., Avants, B., Chiang, M. C., Christensen, G. E., Collins, D. L., Gee, J., Hellier, P., Song, J. H., Jenkinson, M., Lepage, C., Rueckert, D., Thompson, P., Vercauteren, T., Woods, R. P., Mann, J. J. and Parsey, R. V. (2009). Evaluation of 14 nonlinear deformation algorithms applied to human brain MRI registration. *Neuroimage* 46: 786–802.

Klöppel, S., Stonnington, C. M., Chu, C., Draganski, B., Scahill, R. I., Rohrer, J. D., Fox, N. C., Jack, C. R., Ashburner, J. and Frackowiak, R. S. J. (2008). Automatic classification of MR scans in Alzheimer's disease. *Brain* 131: 681–689.

Kober, H. and Wager, T. D. (2010). Meta-analysis of neuroimaging data. *Wiley Interdisciplinary Reviews: Cognitive Science* 1: 293–300.

Kriegeskorte, N., Goebel, R. and Bandettini, P. (2006). Information-based functional brain mapping. *Proceedings of the National Academy of Sciences of the United States of America* 103: 3863–3868.

Laird, A. R., Eickhoff, S. B., Rottschy, C., Bzdok, D., Ray, K. L. and Fox, P. T. (2013). Networks of task co-activations. *Neuroimage* 80: 505–514.

Lange, N., Strother, S. C., Anderson, J. R., Nielsen, F. A., Holmes, A. P., Kolenda, T., Savoy, R. and Hansen, L. K. (1999). Plurality and resemblance in fmri data analysis. *NeuroImage* 10: 282–303.

Lazar, N. (2010). *The Statistical Analysis of Functional MRI Data*. Statistics for Biology and Health. New York: Springer.

LeCun, Y., Bengio, Y. and Hinton, G. (2015). Deep learning. *Nature* 521: 436–444.

Lee, M., Shen, H., Huang, J. Z. and Marron, J. (2010). Biclustering via sparse singular value decomposition. *Biometrics* 66: 1087–1095.

Lee, S., Shen, H., Truong, Y., Lewis, M. and Huang, X. (2011). Independent component analysis involving autocorrelated sources with an application to functional magnetic resonance imaging. *Journal of the American Statistical Association* 106: 1009–1024.

Lerch, J. P., Pruessner, J., Zijdenbos, A. P., Collins, D. L., Teipel, S. J., Hampel, H. and Evans, A. C. (2008). Automated cortical thickness measurements from MRI can accurately separate Alzheimer's patients from normal elderly controls. *Neurobiol. Aging* 29: 23–30.

Lett, T. A., Waller, L., Tost, H., Veer, I. M., Nazeri, A., Erk, S., Brandl, E. J., Charlet, K., Beck, A., Vollstädt-Klein, S. et al. (2017). Cortical surface-based threshold-free cluster enhancement and cortexwise mediation. *Human Brain Mapping* 38: 2795–2807.

Lewis, M., Du, G., Sen, S., Kawaguchi, A., Truong, Y., Lee, S., Mailman, R. and Huang, X. (2011). Differential involvement of striato- and cerebello-thalamo-cortical pathways in tremor- and akinetic/rigid-predominant parkinson's disease. *Neuroscience* 177: 230–239.

Li, H., Nickerson, L. D., Nichols, T. E. and Gao, J.-H. (2017). Comparison of a non-stationary voxelation-corrected cluster-size test with tfce for group-level mri inference. *Human Brain Mapping* 38: 1269–1280.

Li, J., Wang, Z. J., Palmer, S. J. and McKeown, M. J. (2008). Dynamic bayesian network modeling of fmri: A comparison of group-analysis methods. *NeuroImage* 41: 398–407.

Li, R., Chen, K., Fleisher, A. S., Reiman, E. M., Yao, L. and Wu, X. (2011). Large-scale directional connections among multi resting-state neural networks in human brain: A functional mri and bayesian network modeling study. *NeuroImage* 56: 1035–1042.

Li, W., Zhang, S., Liu, C.-C. and Zhou, X. J. (2012). Identifying multi-layer gene regulatory modules from multi-dimensional genomic data. *Bioinformatics* 28: 2458–2466.

Lindquist, M. A. (2008). The statistical analysis of fmri data. *Statistical Science* 23: 439–464.

Liu, J., Li, M., Pan, Y., Lan, W., Zheng, R., Wu, F.-X. and Wang, J. (2017). Complex brain network analysis and its applications to brain disorders: a survey. *Complexity* 2017.

Liu, S., Cai, W., Liu, S., Zhang, F., Fulham, M., Feng, D., Pujol, S. and Kikinis, R. (2015). Multimodal neuroimaging computing: the workflows, methods, and platforms. *Brain Informatics* 2: 181–195.

Lo, C. Y., He, Y. and Lin, C. P. (2011). Graph theoretical analysis of human brain structural networks. *Reviews in the Neurosciences* 22: 551–563.

Magnin, B., Mesrob, L., Kinkingnehun, S., Pelegrini-Issac, M., Colliot, O., Sarazin, M., Dubois, B., Lehericy, S. and Benali, H. (2009). Support vector machine-based classification of Alzheimer's disease from whole-brain anatomical MRI. *Neuroradiology* 51: 73–83.

Mahmoudi, A., Takerkart, S., Regragui, F., Boussaoud, D. and Brovelli, A. (2012). Multivoxel pattern analysis for fmri data: a review. *Computational and Mathematical Methods in Medicine* 2012.

Mandal, P. K., Banerjee, A., Tripathi, M. and Sharma, A. (2018). A comprehensive review of magnetoencephalography (meg) studies for brain functionality in healthy aging and alzheimer's disease (ad). *Frontiers in Computational Neuroscience* 12: 60.

Mateos-Pérez, J. M., Dadar, M., Lacalle-Aurioles, M., Iturria-Medina, Y., Zeighami, Y. and Evans, A. C. (2018). Structural neuroimaging as clinical predictor: A review of machine learning applications. *NeuroImage: Clinical* 20: 506–522.

Miller, M. B., Donovan, C. L., Van Horn, J. D., German, E., Sokol-Hessner, P. and Wolford, G. L. (2009). Unique and persistent individual patterns of brain activity across different memory retrieval tasks. *Neuroimage* 48: 625–635.

Montagna, S., Wager, T., Barrett, L. F., Johnson, T. D. and Nichols, T. E. (2018). Spatial bayesian latent factor regression modeling of coordinate-based metaanalysis data. *Biometrics* 74: 342–353.

Monti, M. M. (2011). Statistical analysis of fmri time-series: A critical review of the glm approach. *Frontiers in Human Neuroscience* 5: 28.

Mumford, J. A. and Nichols, T. (2006). Modeling and inference of multisubject fMRI data. *Engineering in Medicine and Biology Magazine, IEEE* 25: 42–51.

Nathoo, F. S., Kong, L., Zhu, H. and Initiative, A. D. N. (2019). A review of statistical methods in imaging genetics. *Canadian Journal of Statistics* 47: 108–131.

Nichols, T. and Hayasaka, S. (2003). Controlling the familywise error rate in functional neuroimaging: A comparative review. *Statistical Methods in Medical Research* 12: 419–446.

Nichols, T. and Holmes, A. (2002). Nonparametric permutation tests for functional neuroimaging: A primer with examples. *Human Brain Mapping* 15: 1–25.

Nichols, T. E. (2012). Multiple testing corrections, nonparametric methods, and random field theory. *Neuroimage* 62: 811–815.

Oliveira, F. P. and Tavares, J. M. R. (2014). Medical image registration: a review. Computer Methods in Biomechanics and Biomedical Engineering 17: 73–93.

O'Toole, A. J., Jiang, F., Abdi, H., Pénard, N., Dunlop, J. P. and Parent, M. A. (2007). Theoretical, statistical, and practical perspectives on pattern-based classification approaches to the analysis of functional neuroimaging data. Journal of Cognitive Neuroscience 19: 1735–1752.

Owen, A. B. and Perry, P. O. (2009). Bi-cross-validation of the svd and the nonnegative matrix factorization. *The Annals of Applied Statistics*, pp. 564–594.

Pellegrini, E., Ballerini, L., Hernandez, M. d. C. V., Chappell, F. M., González-Castro, V., Anblagan, D., Danso, S., Muñoz-Maniega, S., Job, D., Pernet, C. et al. (2018). Machine learning of neuroimaging for assisted diagnosis of cognitive impairment and dementia: A systematic review. *Alzheimer's & Dementia: Diagnosis, Assessment & Disease Monitoring* 10: 519–535.

Penny, W., Stephan, K., Mechelli, A. and Friston, K. (2004). Modelling functional integration: a comparison of structural equation and dynamic causal models. *NeuroImage* 23: 264–274.

Phan, T. G., Chen, J., Donnan, G., Srikanth, V., Wood, A. and Reutens, D. C. (2010). Development of a new tool to correlate stroke outcome with infarct topography: a proof-of-concept study. *Neuroimage* 49: 127–133.

Platt, J. et al. (1999). Probabilistic outputs for support vector machines and comparisons to regularized likelihood methods. *Advances in Large Margin Classifiers* 10: 61–74.

Poldrack, R., Mumford, J. and Nichols, T. (2011). *Handbook of Functional MRI Data Analysis*. New York: Cambridge University Press.

Polzehl, J. and Tabelow, K. (2019). *Magnetic Resonance Brain Imaging*. Springer.

Popovic, A., de La Fuente, M., Engelhardt, M. and Radermacher, K. (2007). Statistical validation metric for accuracy assessment in medical image segmentation. *International Journal of Computer Assisted Radiology and Surgery* 2: 169–181.

Radua, J. and Mataix-Cols, D. (2009). Voxel-wise meta-analysis of grey matter changes in obsessive–compulsive disorder. *The British Journal of Psychiatry* 195: 393–402.

Radua, J. and Mataix-Cols, D. (2012). Meta-analytic methods for neuroimaging data explained. *Biology of Mood & Anxiety Disorders* 2: 6.

Radua, J., Romeo, M., Mataix-Cols, D. and Fusar-Poli, P. (2013). A general approach for combining voxel-based meta-analyses conducted in different neuroimaging modalities. *Current Medicinal Chemistry* 20: 462–466.

Radua, J., van den Heuvel, O. A., Surguladze, S. and Mataix-Cols, D. (2010). Meta-analytical comparison of voxel-based morphometry studies in obsessive-compulsive disorder vs other anxiety disorders. *Archives of General Psychiatry* 67: 701–711.

Raffelt, D. A., Smith, R. E., Ridgway, G. R., Tournier, J.-D., Vaughan, D. N., Rose, S., Henderson, R. and Connelly, A. (2015). Connectivity-based fixel enhancement: Whole-brain statistical analysis of diffusion mri measures in the presence of crossing fibres. *Neuroimage* 117: 40–55.

Raichle, M. E., MacLeod, A. M., Snyder, A. Z., Powers, W. J., Gusnard, D. A. and Shulman, G. L. (2001). A default mode of brain function. *Proceedings of the National Academy of Sciences* 98: 676–682.

Rajapakse, J. C., Tan, C. L., Zheng, X., Mukhopadhyay, S. and Yang, K. (2006). Exploratory analysis of brain connectivity with ica. *IEEE Eng. Med. Biol. Mag.* 25: 102–11.

Rajapakse, J. C. and Zhou, J. (2007). Learning effective brain connectivity with dynamic Bayesian networks. *Neuroimage* 37: 749–760.

Rathore, S., Habes, M., Iftikhar, M. A., Shacklett, A. and Davatzikos, C. (2017). A review on neuroimaging-based classification studies and associated feature extraction methods for alzheimer's disease and its prodromal stages. *NeuroImage* 155: 530–548.

Reid, A. T., Bzdok, D., Genon, S., Langner, R., Müller, V. I., Eickhoff, C. R., Hoffstaedter, F., Cieslik, E.-C., Fox, P. T., Laird, A. R. et al. (2016). Anima: A data-sharing initiative for neuroimaging meta-analyses. *Neuroimage* 124: 1245–1253.

Reiss, P. T. and Ogden, R. T. (2010a). Functional generalized linear models with images as predictors. *Biometrics* 66: 61–69.

Reiss, P. T. and Ogden, R. T. (2010b). Functional generalized linear models with images as predictors. *Biometrics* 66: 61–69.

Rodriguez, M. Z., Comin, C. H., Casanova, D., Bruno, O. M., Amancio, D. R., Costa, L. d. F. and Rodrigues, F. A. (2019). Clustering algorithms: A comparative approach. *PloS One* 14.

Rubinov, M. and Sporns, O. (2010). Complex network measures of brain connectivity: Uses and interpretations. *NeuroImage* 52: 1059–1069.

Ryali, S., Supekar, K., Abrams, D. A. and Menon, V. (2010). Sparse logistic regression for whole-brain classification of fMRI data. *Neuroimage* 51: 752–764.

Salimi-Khorshidi, G., Smith, S. M., Keltner, J. R., Wager, T. D. and Nichols, T. E. (2009). Meta-analysis of neuroimaging data: a comparison of image-based and coordinate-based pooling of studies. *Neuroimage* 45: 810–823.

Samartsidis, P., Montagna, S., Nichols, T. E. and Johnson, T. D. (2017). The coordinate-based meta-analysis of neuroimaging data. *Statistical Science: A Review Journal of the Institute of Mathematical Statistics* 32: 580.

Sanz-Arigita, E. J., Schoonheim, M. M., Damoiseaux, J. S., Rombouts, S. A., Maris, E., Barkhof, F., Scheltens, P. and Stam, C. J. (2010). Loss of 'smallworld' networks in Alzheimer's disease: graph analysis of FMRI resting-state functional connectivity. *PLoS ONE* 5: e13788.

Schwarzer, G., Carpenter, J. R. and Rücker, G. (2015). *Meta-analysis with R*, volume 4784. Springer.

Sen, S., Kawaguchi, A., Truong, Y., Lewis, M. and Huang, X. (2010). Dynamic changes in cerebello-thalamo-cortical motor circuitry during progression of parkinson's disease. *Neuroscience* 166: 712–719.

Shen, D. and Davatzikos, C. (2002). Hammer: Hierarchical attribute matching mechanism for elastic registration. *IEEE Trans. Med. Imaging* 21: 1421–1439.

Shen, H. and Huang, J. Z. (2008). Sparse principal component analysis via regularized low rank matrix approximation. *Journal of Multivariate Analysis* 99: 1015–1034.

Skup, M. (2010). Longitudinal fMRI analysis: A review of methods. *Stat Interface* 3: 232–252.

Smith, S. M., Jenkinson, M., Woolrich, M. W., Beckmann, C. F., Behrens, T. E. J., Johansen-berg, H., Bannister, P. R., Luca, M. D., Drobnjak, I., Flitney, D. E., Niazy, R. K., Saunders, J., Vickers, J., Zhang, Y., Stefano, N. D., Brady, J. M. and Matthews, P. M. (2004). Advances in functional and structural mr image analysis and implementation as fsl. *NeuroImage* 23: 208–219.

Smith, S. M., Miller, K. L., Salimi-Khorshidi, G.,Webster, M., Beckmann, C. F., Nichols, T. E., Ramsey, J. D. and Woolrich, M.W. (2011). Network modelling methods for fmri. *NeuroImage* 54: 875–891.

Smith, S. M. and Nichols, T. E. (2009). Threshold-free cluster enhancement: addressing problems of smoothing, threshold dependence and localisation in cluster inference. *Neuroimage* 44: 83–98.

Stephan, K. E. and Friston, K. J. (2010). Analyzing effective connectivity with fMRI. *Wiley Interdiscip. Rev. Cogn. Sci.* 1: 446–459.

Stone, J. V., Porrill, J., Porter, N. R. and Wilkinson, I. D. (2002). Spatiotemporal independent component analysis of event-related fMRI data using skewed probability density functions. *Neuroimage* 15: 407–421.

Storey, J. (2002). A direct approach to false discovery rates. *Journal of the Royal Statistical Society: Series B (Statistical Methodology)* 64: 479–498.

Storey, J. (2003). The positive false discovery rate: A bayesian interpretation and the q-value. *Annals of Statistics* 31: 2013–2035.

Suk, H.-I., Lee, S.-W., Shen, D., Initiative, A. D. N. et al. (2017). Deep ensemble learning of sparse regression models for brain disease diagnosis. *Medical Image Analysis* 37: 101–113.

Sun, F. T., Miller, L. M. and D'Esposito, M. (2004). Measuring interregional functional connectivity using coherence and partial coherence analyses of fMRI data. *Neuroimage* 21: 647–658.

Taniwaki, T., Okayama, A., Yoshiura, T., Togao, O., Nakamura, Y., Yamasaki, T., Ogata, K., Shigeto, H., Ohyagi, Y., Kira, J.-i. et al. (2007). Age-related alterations of the functional interactions within the basal ganglia and cerebellar motor loops *in vivo*. *Neuroimage* 36: 1263–1276.

Tavor, I., Jones, O. P., Mars, R., Smith, S., Behrens, T. and Jbabdi, S. (2016). Task-free mri predicts individual differences in brain activity during task performance. *Science* 352: 216–220.

Teipel, S. J., Born, C., Ewers, M., Bokde, A. L., Reiser, M. F., Moller, H. J. and Hampel, H. (2007). Multivariate deformation-based analysis of brain atrophy to predict Alzheimer's disease in mild cognitive impairment. *Neuroimage* 38: 13–24.

Tian, T. S. (2010). Functional data analysis in brain imaging studies. *Frontiers in Psychology* 1: 35.

Turkeltaub, P. E., Eden, G. F., Jones, K. M. and Zeffiro, T. A. (2002). Metaanalysis of the functional neuroanatomy of single-word reading: method and validation. *Neuroimage* 16: 765–780.

Turkeltaub, P. E., Eickhoff, S. B., Laird, A. R., Fox, M., Wiener, M. and Fox, P. (2012). Minimizing within-experiment and within-group effects in activation likelihood estimation meta-analyses. *Human Brain Mapping* 33: 1–13.

van den Heuvel, M. P. and Pol, H. E. H. (2010). Exploring the brain network: A review on resting-state fmri functional connectivity. *European Neuropsychopharmacology* 20: 519–534.

Van Essen, D. C., Smith, J., Glasser, M. F., Elam, J., Donahue, C. J., Dierker, D. L., Reid, E. K., Coalson, T. and Harwell, J. (2017). The brain analysis library of spatial maps and atlases (balsa) database. *Neuroimage* 144: 270–274.

Vapnik, V. (1999). *The Nature of Statistical Learning Theory*. Statistics for Engineering and Information Science. New York: Springer.

Vemuri, P., Gunter, J. L., Senjem, M. L., Whitwell, J. L., Kantarci, K., Knopman, D. S., Boeve, B. F., Petersen, R. C. and Jack, C. R. (2008). Alzheimer's disease diagnosis in individual subjects using structural MR images: validation studies. *NeuroImage* 39: 1186–97.

Vieira, S., Pinaya, W. H. and Mechelli, A. (2017). Using deep learning to investigate the neuroimaging correlates of psychiatric and neurological disorders: Methods and applications. *Neuroscience & Biobehavioral Reviews* 74: 58–75.

Viviani, R., Gron, G. and Spitzer, M. (2005). Functional principal component analysis of fMRI data. *Human Brain Mapping* 24: 109–129.

Wager, T. D., Jonides, J. and Reading, S. (2004). Neuroimaging studies of shifting attention: a meta-analysis. *Neuroimage* 22: 1679–1693.

Wager, T. D., Lindquist, M. and Kaplan, L. (2007). Meta-analysis of functional neuroimaging data: current and future directions. Social Cognitive and Affective Neuroscience 2: 150–158.

Warnick, R., Guindani, M., Erhardt, E., Allen, E., Calhoun, V. and Vannucci, M. (2018). A bayesian approach for estimating dynamic functional network connectivity in fmri data. *Journal of the American Statistical Association* 113: 134–151.

Watts, D. J. and Strogatz, S. H. (1998). Collective dynamics of 'small-world' networks. *Nature* 393: 440–442.

Weygandt, M., Blecker, C. R., Schafer, A., Hackmack, K., Haynes, J. D., Vaitl, D., Stark, R. and Schienle, A. (2012). fMRI pattern recognition in obsessive-compulsive disorder. *Neuroimage* 60: 1186–1193.

Weygandt, M., Schaefer, A., Schienle, A. and Haynes, J. D. (2012). Diagnosing different binge-eating disorders based on reward-related brain activation patterns. *Human Brain Mapping* 33: 2135–2146.

Witten, D. M., Tibshirani, R. and Hastie, T. (2009). A penalized matrix decomposition, with applications to sparse principal components and canonical correlation analysis. *Biostatistics* 10: 515–534.

Wood, S. N. (2011). Fast stable restricted maximum likelihood and marginal likelihood estimation of semiparametric generalized linear models. *Journal of the Royal Statistical Society: Series B (Statistical Methodology)* 73: 3–36.

Woolrich, M. W., Jbabdi, S., Patenaude, B., Chappell, M., Makni, S., Behrens, T., Beckmann, C., Jenkinson, M. and Smith, S. M. (2009). Bayesian analysis of neuroimaging data in fsl. *NeuroImage* 45: S173–86.

Worsley, K. and Taylor, J. (2006). Detecting fmri activation allowing for unknown latency of the hemodynamic response. *NeuroImage* 29: 649–654.

Worsley, K. J. (2003). Detecting activation in fMRI data. *Statistical Methods in Medical Research* 12: 401–418.

Worsley, K. J., Marrett, S., Neelin, P., Vandal, A. C., Friston, K. J., and Evans, A. C. (1996). A unified statistical approach for determining significant signals in images of cerebral activation. *Human brain mapping* 4, 58–73.

Worsley, K. J., Taylor, J. E., Tomaiuolo, F. and Lerch, J. (2004). Unified univariate and multivariate random field theory. *Neuroimage* 23: S189–S195.

Wu, X., Li, R., Fleisher, A. S., Reiman, E. M., Guan, X., Zhang, Y., Chen, K. and Yao, L. (2011). Altered default mode network connectivity in Alzheimer's disease–a resting functional MRI and Bayesian network study. *Human Brain Mapping* 32: 1868–1881.

Xue, W., Kang, J., Bowman, F. D., Wager, T. D. and Guo, J. (2014). Identifying functional co-activation patterns in neuroimaging studies via poisson graphical models. *Biometrics* 70: 812–822.

Yamashita, O., Sato, M. A., Yoshioka, T., Tong, F. and Kamitani, Y. (2008). Sparse estimation automatically selects voxels relevant for the decoding of fMRI activity patterns. *Neuroimage* 42: 1414–1429.

Yoshida, H., Kawaguchi, A., Yamashita, F. and Tsuruya, K. (2018). The utility of a network–based clustering method for dimension reduction of imaging and non-imaging biomarkers predictive of alzheimerfs disease. *Scientific Reports* 8: 1–10.

Yotter, R. A., Nenadic, I., Ziegler, G., Thompson, P. M. and Gaser, C. (2011). Local cortical surface complexity maps from spherical harmonic reconstructions. *NeuroImage* 56: 961–973.

Yu, F., Barron, D. S., Tantiwongkosi, B. and Fox, P. (2015). Patterns of gray matter atrophy in atypical parkinsonism syndromes: a vbm meta-analysis. Brain and Behavior 5: e00329.

Zhang, C.-H. et al. (2010). Nearly unbiased variable selection under minimax concave penalty. *The Annals of Statistics* 38: 894–942.

Zhang, D. and Raichle, M. E. (2010). Disease and the brain's dark energy. *Nature Reviews Neurology* 6: 15–28.

Zheng, X. and Rajapakse, J. C. (2006). Learning functional structure from fMR images. *Neuroimage* 31: 1601–1613.

Ziegler, G., Dahnke, R. and Gaser, C. (2012). Models of the aging brain structure and individual decline. *Front Neuroinform* 6: 3.

Zipunnikov, V., Caffo, B., Yousem, D. M., Davatzikos, C., Schwartz, B. S. and Crainiceanu, C. (2011). Functional principal component model for highdimensional brain imaging. *Neuroimage* 58: 772–784.

Zou, H., Hastie, T. and Tibshirani, R. (2006). Sparse principal component analysis. *Journal of Computational and Graphical Statistics* 15: 265–286.

Index

Milton Keynes UK
Ingram Content Group UK Ltd.
UKHW051015071024
449327UK00012B/262

9 780367 752217